Use Your Fingers,
Use Your Toes

A Capital Ideas book, practical books that offer expert advice on key personal and professional aspects of life. Other titles include:

The R.A.T. (Real-World Aptitude Test) by Homer E. Moyer, Jr.

"Just Sign Here, Honey" by Marilyn Barrett

Our Money Ourselves for Couples by C. Diane Ealy, Ph.D., and Kay Lesh, Ph.D.

Safe Living in a Dangerous World by Nancy Harvey Steorts

The Parent's Guide to Business Travel by Charlie Hudson

The Man Who Would Be Dad by Hogan Hilling

A Grandmother's Guide to Extended Babysitting by Ruth A. Brown

How to Avoid the Mommy Trap by Julie Shields

Father's Milk by Andre Stein and Peter Samu

The Golden Rules of Parenting by Rita Boothby

Graduate by Kristen Gustafson

A Grammar Book for You and I . . . Oops, Me by C. Edward Good

The Kitchen Answer Book by Hank Rubin

Your New Dog by Susan McCullough

The Dogs Who Grew Me by Ann Pregosin

Use Your Fingers, Use Your Toes

Quick and Easy Step-by-Step Solutions to Your Everyday Math Problems

Beth Norcross

A Capital Ideas Book

CAPITAL
BOOKS, INC.

Sterling, Virginia

Capital Books, Inc.
P.O. Box 605
Herndon, Virginia 20172-0605

Talia Greenberg, Illustrator

Library of Congress Cataloging-in-Publication Data

Norcross, Beth.
    Use your fingers, use your toes : quick and easy step-by-step
solutions to your everyday math problems / Beth Norcross.—1st ed.
        p. cm. — (A Capital ideas book)
Includes index.
    ISBN 1-931868-14-X (alk. paper)
    1. Mathematics—Popular books.   I. Title.   II. Series.
QA93.N56 2003
510—dc21

                                                            2003004523

Printed in the United States of America on acid-free paper that meets the American National Standards Institute Z39-48 Standard.

First Edition

10 9 8 7 6 5 4 3 2 1

To Clint, Charlie, Erin, and James,

who continue to lift me up as

well as keep my feet on the ground.

Contents

Acknowledgments

A book such as this is the result of a lifetime of experience, and for that reason it is difficult to acknowledge all who played a part. The inspiration for this book is the many men and women, some I know and others I do not, who struggle day after day with doubling a recipe, helping their children with fractions, changing kilometers to miles, or figuring a percentage. A certain friend, having filled her driveway with a mountain of mulch because she could not figure how much to order for her garden, finally got me off my duff and on to the word processor. These people were my muses, and I thank them.

Charlene Carroll, Stan Fixico, Dick Miller, Laura Peebles, and Jim Warns lent their specialized expertise to review of specific ideas or chapters, and in so doing made the book clearer and more useful. Linda Crill introduced me to the wonderful people at Capital Books, who encouraged this project from the beginning. Particular thanks go to my editors, Noemi Taylor, Judy Karpinski, and Judy Coughlin who helped me turn a number of interesting ideas into what I hope will be a practical and helpful guide. Even mathematicians need math help from time to time, so Gaye Lindsey patiently reviewed every page of the book for mathematical accuracy.

My friend Rebecca Leet saw the potential and practicality of this book when it was just a few random thoughts, and she continued to believe in it and me throughout the long production process. My sister, Susan Fixico, provided wisdom and laughter at all the right moments.

My children, Charlie, James, and Erin, made many sacrifices during the writing of the manuscript, including having to give up their normally healthy eating preferences for pizza and fast food. Finally, and most importantly, I thank my husband, Clint, who read and edited this manuscript countless times and whose enduring patience and kindness sustained me through this endeavor.

Introduction

Everyone has one: a nightmarish memory of some math class that made you feel like a pile of elephant droppings because you just didn't "get" it.

Maybe you were told that you just weren't good at math, or weren't trying hard enough, or were just being stubborn. Well, join the club. Almost everyone struggles with some aspect of math. I have a degree in mathematics, and geometry still gives me a stomachache.

Here's a secret no one ever told you: chances are, if you didn't "get" math, it was because the teacher simply didn't explain it to you in a way you could understand. And you probably know a whole lot more math than you think you do. If only you could remember it when you need it. Like, before the nursery delivers three times more mulch than you need, or before you double the recipe and end up with cookies your child is too embarrassed to take to the school fair.

Now all you'll have to remember is where you stashed this little book. Because inside these pages is all the help you need to handle those pesky everyday math challenges.

What's in This Book

In this book you will find easy-to-understand, step-by-step solutions to your most common math problems.

Part 1, "Math at Home," begins with explaining how to conquer the many math challenges in the kitchen, including adjusting recipes, converting units of measure, and calculating cooking times. This section covers how basic math can help you lose weight by counting calories, computing your personal "calorie deficit," and reading complicated food labels. It also tackles those daunting "how much do I need" questions we all have when doing home improvement projects—like, How much carpeting, paint, or mulch do I need and how much will it cost?

Part 2, "Math Away from Home," takes you through those math problems you encounter at the store, in a restaurant, in a car, and even in the ballpark. It explains simple tipping techniques and how to use math to shop wisely. You will learn how to calculate and understand those complicated percentage markdowns and do effective comparison shopping. This section shows you how to compute gas mileage, calculate travel time, and convert foreign currency. And, for those of you who are curious about what all those numbers in your sports program mean, this part shows you how to understand and calculate basic sports statistics, such as earned run average, shooting and passing percentages, and batting averages.

Part 3, "Personal Finance Math," gently guides you through basic financial calculations. It gives you essential information on your salary and wages, including how to calculate raises and how to understand tax basics. You will learn how to keep a checkbook and to reconcile your bank statement. Part 3 also goes over the basics of making sound investment decisions, starting with an explanation of simple and compound interest. It shows you how to calculate monthly payments and interest on loans, including car loans and mortgages, and it has a whole section devoted to that subject we all love to hate—credit cards.

The appendixes review basic math concepts and calculations useful in solving everyday math problems in this book and beyond it. They include "Twelve Tips to Stress-Free Math" and demonstrate how to work with fractions and percents. The appendixes also cover how to round off and how to calculate perimeter, area, and volume. The final appendix has an easy-to-use chart that walks you through converting units of measure.

Icons Used in This Book:

 Remember reminds you of a basic math concept you might have forgotten. More basic math information can be found in the appendixes.

 Tip gives a hint, suggestion, recommendation, or reminder that is helpful in solving your problem.

 Warning warns of some hidden danger in the solution of this kind of problem.

How to Use This Book

Ultimately, you will use this book like any other reference book—only when you need it. But first, you will want to read the book and see what's here. You may not encounter any one of these problems on a regular basis, but when you do, you will know where to look. After that, put it in your kitchen or on your desk, and then forget about it until you need it. When you encounter a math problem, just look in the table of contents or the index, and go to the relevant section.

The book is written for people with a variety of learning styles. If you like to work methodically through a problem, use the step-by-step examples to help you. If you just need a reminder of how to approach a problem, or you're a "big picture" kind of person, the "In a Nutshell" Boxes are for you. These summarize the information on a specific topic and, when necessary, provide formulas.

All the chapters assume that you use a calculator for addition, subtraction, multiplication, and division. I have a friend who still does all the calculations for her checkbook with a pencil and a piece of paper "just because it's fun." This book is not for her.

The examples in the book are chosen because they happen in

real life. Therefore, the solutions aren't always pretty; rounding is used liberally to get answers into a more user-friendly form—just like in real life. So, if you are following along with a particular example, and you don't get exactly the same answer, don't worry. It's probably due to the rounding. When you need to know how to round in specific instances, the book goes over that, and there is a detailed explanation of rounding in appendix 4.

Math Do's and Don'ts

Before you start reading the book, here are a few basic suggestions on how to approach an everyday math problem.

1. Don't—panic. Life's too short.
2. Do—take a deep breath. It won't be that bad, I promise.
3. Don't—let anyone tell you there is just one way to solve a problem. It's not true.
4. Do—start. Start in the middle of a problem, at the end, anywhere, just start.
5. Don't—show your work. It's nobody's business but yours.
6. Do—guess. Use your imagination. Use your intuition. Use your fingers. Use your toes.

Now, just relax, knowing you will never again have to worry about those frustrating everyday math problems. Help is on the way.

1 Math at Home

1 Food Preparation

E ven the simplest kitchen task quickly becomes complicated. "Cook the beef for 18 to 20 minutes per pound." Why isn't there a cookbook that simply tells you how long to cook a six-pound roast? A recipe that serves six and a guest list that includes an army of fifteen—what are you supposed to do with that? Your recipe calls for half a cup of heavy cream, but at the store cream is sold only in pints. How much do you buy?

Not to worry. We'll do it all here. You will learn how long to cook meat, how much meat you need, how many pies to make, and how much food you need to feed that army or just the two of you. We'll double recipes, halve recipes, and even one-and-one-half-times a recipe. In addition, we'll deal with recipes that call for a change from one kind of measure to another.

How Much Food Do You Need?

The most common math problem associated with preparing food is figuring out how much food you need to make. A recipe might serve six, but you expect eight people for dinner. You know a pie has eight slices, but your family numbers ten. How do you figure out how much to make? The following example will walk you through how to calculate the volume of meat, pies (including pizzas), and other kinds of food you need to prepare.

PROBLEM

It's Thanksgiving. Your mother-in-law has declared with great fanfare that she is going to "let" you cook the traditional meal for the entire extended family this year. She wants you to be sure to include her traditional recipe for goulash as well as her son's favorite dessert, mince pie. Twenty people to serve and you don't know where to start.

Meats (poultry, fish, or other food measured by the pound)

Okay, let's start with the turkey, and I don't mean you-know-who. The label on Tom's Terrific Turkeys says that an average serving size is ¾ pound.

STEP 1

Determine an average serving size. In this example: ¾ **pound per person.**

 Most food labels will suggest an average serving size. You can also ask your butcher or consult a cookbook.

STEP 2

Multiply the serving size (¾ pound) by the number of people you are serving (20).

.75 pounds times 20 people = 15 pounds

(Change ¾ to .75 so you can use your calculator.)

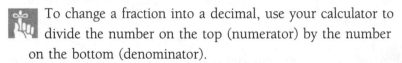 To change a fraction into a decimal, use your calculator to divide the number on the top (numerator) by the number on the bottom (denominator).

 The serving size might be in ounces instead of pounds. The turkey label above could have said 12 ounces per person instead of ¾ pound per person. In that case, you would just multiply the serving size in ounces times the number of people (12 ounces times 20 people = 240 ounces). Then change ounces to pounds by dividing by 16, since there are 16 ounces in a pound (240 ounces divided by 16 = 15 pounds).

SOLUTION

You need a 15-pound turkey.

(Buy a somewhat bigger turkey if you want leftovers or if you are expecting cousin Bruno.)

Pies (Cakes and Pizzas, Too)

PROBLEM

Now let's figure out how many mince pies you will need to make for dessert. This method also works for ordering pizzas.

STEP 1

Determine how many slices of pie each person will eat. Here, you use 1½, since some will eat one slice and some will eat two.

1½ slices (or 1.5) slices each

(Change 1½ to 1.5 so you can use your calculator more easily for the next steps. Remember that 1½ = 1 plus ½. To make ½ into a decimal, use your calculator to divide the top number by the bottom number. The result is .5. Substituting .5 for ½, 1½ becomes 1.0 plus .5, which equals 1.5.)

STEP 2

Multiply the number of pie slices each person will eat (1.5) by the number of people expected (20).

1.5 times 20 = 30 slices of pie

STEP 3

Determine the number of slices per pie—typically it is eight.

8

STEP 4

Divide the total pie slices needed (30) by the number of slices per pie (8).

30 divided by 8 = 3.75 pies

SOLUTION

Round up, and make (or order) 4 pies.

Other Dishes (Recipes)

PROBLEM

Now it is time to figure out how much to increase your mother-in-law's famous goulash recipe (or any other recipe you need to make).

STEP 1

Determine how many people the recipe serves. In the case of your mother-in-law's famous goulash, it serves six.

6 people

STEP 2

Divide the total number of people expected (20) by the number of servings in the recipe (6).

20 divided by 6 = 3.33 recipes

SOLUTION

You will need 3 to 4 times the amount the recipe calls for.
(To calculate amounts for each ingredient, see Changing Recipes below.)

HOW MUCH FOOD DO YOU NEED?

Meats (poultry, fish, or other food measured by the pound)

Pounds of meat needed = number of people expected × pounds per average serving

Pies (cakes and pizzas, too)

Slices needed = slices each will eat × number of people

Number of pies needed = slices needed ÷ total number of slices in one pie

Other Dishes (recipes)

Recipe multiplier = total number of people ÷ number of people the recipe serves

Amount of each ingredient needed = ingredients in recipe × the recipe multiplier

Changing Recipes

Now that you have figured out how much food you need, you will often have to increase or decrease a recipe to make an appropriate amount of food. The following examples will walk you through how to increase and decrease a recipe. As a fair warning, you should know that there are lots of fractions involved in changing recipes. We will step through it here, but if you need extra review on adding, multiplying, or dividing fractions, see appendix 2.

Increasing a Recipe

PROBLEM

You are about to prepare dinner, and your son calls to ask if he can bring a "few" friends to dinner—5, to be

exact. You agree, and then realize that your Cluckin' Good Chicken recipe feeds the 6 of you in the family, but will not be enough to feed Bud's 5 friends, too. You need to increase the recipe, but how?

Start with the list of ingredients for the recipe that serves 6 people:

Ingredients for Cluckin' Good Chicken

6 medium chicken breasts

½ cup chopped onion

¼ cup celery

½ teaspoon salt

1 can mushroom soup

⅔ cup sour cream

STEP 1

Determine the total number of people you'll be feeding. Add a couple more if the people are big eaters.

6 people plus 5 people = 11 people

STEP 2

Determine how many the recipe serves.

6 people

STEP 3

Divide the total number of people who are eating (11) by the number the recipe serves (6).

11 divided by 6 = 1.8

STEP 4

Round up to the nearest whole number to get the recipe multiplier.

2

STEP 5

Multiply all the ingredients by the multiplier, 2.

SOLUTION

6 chicken breasts times 2 = 12 chicken breasts

½ cup onion times 2 = two ½ cups = 1 cup onion

¼ cup celery times 2 = two ¼ cups = ½ cup celery

½ teaspoon salt times 2 = two ½ teaspoons = 1 teaspoon salt

1 can of soup times 2 = 2 cans of soup

⅔ cup sour cream times 2 = ⁴⁄₃ cup = or 1⅓ cup sour cream

To multiply a fraction by a number, multiply the top of the fraction (numerator) by the number and leave the bottom (denominator) of the fraction alone. (⅔ times 2 = ⁴⁄₃. You have two choices for what to do next. The easiest thing is to measure out ⅓ cup four times. Or, you can change ⁴⁄₃ into a "mixed number" [a whole number and a fraction combined]. To change ⁴⁄₃ to a mixed number, first divide the top by the bottom without your calculator. 4 divided by 3 = 1 with 1 leftover. The first number—1—is the whole number. Now stick the leftover, coincidentally also 1, over the denominator—3—to make the fraction part of your mixed number [⅓]. Now put them together to get 1⅓. Not too bad, was it?)

One-and-a-Half-ing a Recipe

PROBLEM

You were about to tackle your recipe for Cluckin' Good Chicken when your son, Bud, called to say that 3 of his buddies can't make it for dinner after all. Now there will

be only 2 of his buddies coming, making a total of 8 people. You use the steps above to figure the multiplier is about 1.3. A single recipe for 6 will not be enough, but doubling is too much. So, you round up the 1.3 to 1.5, which amounts to making 1½ times the recipe. How do you do that?

STEP 1

Start by figuring out what half the recipe would be by dividing all the original ingredient quantities by 2.

6 chicken breasts divided by 2 = 3 chicken breasts

½ cup onion divided by 2 = ¼ cup onion (since there are two ¼ cups in ½ cup)

¼ cup celery divided by 2 = ⅛ cup celery (since there are two ⅛ cups in ¼ cup)

½ teaspoon salt divided by 2 = ¼ teaspoon salt (since there are two ¼ teaspoons in ½ teaspoon)

1 can of soup divided by 2 = ½ can of soup

⅔ cup sour cream divided by 2 = ⅓ cup sour cream

When dividing fractions by a number, just leave the top (numerator) alone, and multiply the bottom (denominator) by the number you are dividing by. For example, ⅓ divided by 2 = ⅙.

STEP 2

Add the original quantity of ingredients to the ½ quantity of ingredients from Step 1.

6 chicken breasts plus 3 chicken breasts = 9 chicken breasts

½ cup onion plus ¼ cup onion = ²⁄₄ cup onion plus ¼ cup onion = ¾ cup onion

To add fractions with different bottoms (denominators) you have to change one or both of the fractions so that the bottoms of both fractions are the same. To do this, just multiply the top and bottom of one of the fractions by the same number until you get the bottoms to look alike. It doesn't change the value of the fraction at all. In this case, change ½ into ²⁄₄ cup by multiplying the top and bottom of ½ by 2: $\dfrac{1 \text{ times } 2 = 2}{2 \text{ times } 2 = 4}$. So, ½ onion = ²⁄₄ onion. Now that the bottoms look the same, you can add the fractions. Do the same for the other ingredients where necessary.

¼ (²⁄₈) cup celery plus ⅛ cup celery = ³⁄₈ cup celery

½ (²⁄₄) teaspoon salt plus ¼ teaspoon salt = ¾ teaspoon salt

1 can of soup plus ½ can of soup = 1½ cans of soup

⅔ cup sour cream plus ⅓ cup sour cream = ³⁄₃ or 1 cup sour cream

When you're adding fractions, like ½ and ¼, and can't remember how to get them into the same form, just use the ½ cup measure, and the ¼ cup measure, and get on with life!

SOLUTION

You will need the following ingredients:

9 chicken breasts

¾ cup chopped onion

³⁄₈ cup celery (use just use a little more than ¼ cup or a little less than ½ cup)

¾ teaspoon salt

1½ cans mushroom soup

1 cup sour cream

Decreasing a Recipe

PROBLEM

You are all ready to go with Cluckin' Good Chicken when the phone rings—again. Your son, Bud, and his pals are going out for a burger instead of coming to your house for dinner. Your wife is working late and will grab something at the office, and your teenage daughter has just stormed through the kitchen announcing she has become a vegetarian and will not eat (yuck!) chicken. Okay, you resist the urge to just give up and get pizza delivered for the 3 of you who are left. Instead, you will try to cut this recipe in half (since 3 is ½ of 6, the number of poeple the recipe serves). What do you do now?

STEP 1

To halve the recipe, divide all the original ingredient quantities by 2.

6 chicken breasts divided by 2 = 3 chicken breasts

½ cup onion divided by 2 = ¼ cup onion

To divide fractions, *multiply* the bottom (denominator) by the number you are dividing by, and leave the top (numerator) as is. For example, ½ divided by 2 is ¼, since you leave the 1 alone and multiply the bottom (2) by 2 to make 4.

¼ cup celery divided by 2 = ⅛ cup celery

½ teaspoon salt divided by 2 = ¼ teaspoon salt

1 can of soup divided by 2 = ½ can of soup

⅔ cup sour cream divided by 2 = ⅓ cup sour cream (since there are two ⅓ cups in ⅔ cup)

 You might know how many units make up a larger measure without math calculations. For example, you know intuitively there are two ⅓s in ⅔; two ¼s in ½; three ¼s in ¾; and so on. Trust your instincts, and do it whatever way makes sense to you.

SOLUTION

You will need the following ingredients:

3 chicken breasts

¼ cup chopped onion

⅛ cup celery

(If you don't have a ⅛-cup measure, estimate half of a ¼-cup measure.)

¼ teaspoon salt

½ can mushroom soup

⅓ cup sour cream

In a Nutshell

CHANGING RECIPES

Increasing a Recipe

Recipe multiplier – total number of people to serve ÷ number of servings in the recipe (round to nearest .5)

Ingredients needed = ingredients in recipe × recipe multiplier

One-and-a-Half-ing a Recipe

Ingredients needed = ingredients in recipe + (ingredients in recipe ÷ 2)

Decreasing a Recipe

Ingredients needed for ½ the recipe = ingredients in recipe ÷ 2

(Divide by 3 for ⅓ the recipe, 4 for ¼, etc.)

 Be careful when increasing and decreasing salt and other spices. For reasons far beyond my culinary knowledge, they

do not change proportionately. Better taste as you go. In addition, it is difficult to increase some recipes, particularly those that make dough. Check the recipe and your cookbooks carefully for information about these situations.

Changing from One Measure to Another

Often in cooking food and buying ingredients, you will be faced with having to change from one kind of measure to another, from tablespoons to teaspoons, quarts to gallons, and so forth. It is hard to remember all those equivalents, like eight ounces in a cup and four quarts in a gallon, and even harder to remember when it is appropriate to multiply or divide. Relax. It is all laid out in the easy Food Preparation Changing Table below. Just follow the simple steps, and you'll never have to worry about different measurements again.

Changing Table
FOOD PREPARATION

To go from . . .	to . . .	you . . .	. . . . .	by . . .
tablespoons	teaspoons	multiply	# of tablespoons	3
teaspoons	tablespoons	divide	# of teaspoons	3
¼ cups	tablespoons	multiply	# of ¼ cups	4
tablespoons	¼-cups	divide	# of tablespoons	4
cups	ounces	multiply	# of cups	8
ounces	cups	divide	# of ounces	8
pints	cups	multiply	# of pints	2
cups	pints	divide	# of cups	2
quarts	cups	multiply	# of quarts	4
cups	quarts	divide	# of cups	4
quarts	pints	multiply	# of quarts	2
pints	quarts	divide	# of pints	2
gallons	quarts	multiply	# of gallons	4
quarts	gallons	divide	# of quarts	4

The next couple of examples will lead you through how to use the table.

Cups to Pints

PROBLEM

Your husband runs into his old high school girlfriend and her husband at the mall. Somewhere between telling you how she looks just the same and has just published her third novel, he tells you he has invited them over for dinner Saturday night. You're not the insecure type, but nonetheless decide to tackle a complicated, but impressive, parfait for dessert. The recipe calls for 3 cups of whipping cream, but you recall that cream comes in containers measured in pints, not cups. How many pints of whipping cream do you need?

STEP 1

Find the line in the Food Preparation Changing Table on page 14 that tells you how to convert **cups** to **pints**.

To go from . . .	to . . .	you . . .	. . . .	by . . .
cups	pints	divide	# of cups	2

STEP 2

Following the instructions on the Food Preparation Changing Table on page 14, divide 3 cups by 2.

3 cups divided by 2 = 1½ pints

SOLUTION

You need to buy 1½ pints of whipping cream for the 3 cups the recipe calls for.

Tablespoons to Teaspoons

PROBLEM

The parfait has turned out perfectly, and you're just finishing up the green beans amandine. You have halved all the ingredients in the recipe for eight successfully except the one tablespoon of butter. You don't have a ½-tablespoon measure, and you know you're supposed to change tablespoons to teaspoons, but you don't have a clue how to do this.

STEP 1

Find the line in the Food Preparation Changing Table on page 14 that tells you how to go from **tablespoons** to **teaspoons**.

To go from . . .	to . . .	you . . .	. . . .	by . . .
tablespoons	teaspoons	multiply	# of table-spoons	3

STEP 2

Following the instructions on the Food Preparation Changing Table on page 14, multiply ½ tablespoon by 3.

½ times 3 = ³⁄₂ or 1½ teaspoons

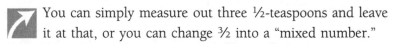 You can simply measure out three ½-teaspoons and leave it at that, or you can change ³⁄₂ into a "mixed number."

A mixed number is a number with a whole number and a fraction. To figure out the mixed number for ³⁄₂, first divide the top by the bottom without your calculator: 3 divided by 2 = 1 with a remainder of 1. The first number—1—is the whole number. Now place the remainder, coincidentally also 1, over the bottom of your fraction—2—to form ½. Put them together to get 1½ teaspoon.

SOLUTION

Use 1½ teaspoons of butter.

Calculating Cooking Time

Figuring out the hours and minutes to cook a roast or other food with "minutes per pound" cooking instructions can be tricky. With the Cooking Time Table below and a few quick calculations, it will be easy.

Cooking Time Table		
Minutes		Hours
60	in	1
120	in	2
180	in	3
240	in	4
300	in	5
360	in	6

Look at the following example to see how the table works.

PROBLEM

Your tenth anniversary is here. The kids are with Grandma and Grandpa, doing you don't care what. The table is set with fine china, silver, and candlesticks you haven't used since the kids were born. You have prepared a fabulous meal, including a crown roast. All you have to do is throw it in the oven. The recipe says to cook it for 25 to 30 minutes per pound. Yikes. What does that mean? You just want to know what to set the buzzer for.

STEP 1

Rummage through the garbage and find the roast wrapper and the label with the total weight and suggested cooking times.

5.7 pounds

25 to 30 minutes per pound

STEP 2

To calculate cooking time, multiply the weight (5.7) by the lesser amount of cooking time per pound (25). (You can always cook it longer if it's not done.)

5.7 times 25 = 142.5 minutes

(Round to 143 minutes)

You could divide 143 minutes by the 60 minutes in an hour to come up with 2.38 hours of cooking time. However, that doesn't mean the cooking time is 2 hours and 38 minutes. It's actually 2 hours plus .38 of an hour. It will be easier to follow Step 3 instead, which uses the simple Cooking Time Table.

STEP 3

To find cooking time in hours and minutes, use the Cooking Time Table on page 17. First, find the highest number in the "Minutes" column *that does not exceed your total cooking time.* Also note the number in the "Hours" column corresponding with those minutes. This number will be the number of hours you should cook the meat.

120 minutes in 2 hours

STEP 4

Subtract the number of minutes found in Step 3 (120) from your total cooking time in Step 2 (143).

143 minus 120 = 23 minutes

SOLUTION

Add the hours of cooking time from Step 3 (2) and the minutes from Step 4 (23).

Cook the roast for 2 hours and 23 minutes.

Cooking times are always approximate. Be sure to check the roast for doneness well before the total cooking time has elapsed.

In a Nutshell

COOKING TIME

1. Calculate total cooking time in minutes by multiplying how many pounds of meat you are cooking by the shortest cooking time per pound suggested on the label.

2. Find the number of minutes closest to, but not exceeding, your total cooking time in the "Minutes" column of the Cooking Time Table (page 17).

3. Note the number of hours on the table that corresponds with the "Minutes" number from #2. This is the total number of hours you will cook the meat.

4. To find the additional minutes of cooking time, subtract the "Minutes" number in #2 from the total cooking time minutes in #1.

5. Combine the hours from #3 and the minutes from #4 to get the final cooking time in hours and minutes.

Armed with all the basic math you need to cook up a storm, you could feed an army, or at least a small regiment. Julia Child would be so proud.

2 Weight Loss and Exercise

Wait a minute. Isn't this a book about everyday math? What's math got to do with weight loss? Everything, really. No matter what eating plan you are following, no matter how much you weigh, no matter how much exercise you get, weight loss is all about the number of calories you take in by eating and the number of calories you use up by moving around.

If you take in about the same number of calories you use up, your weight stays about the same. If you take in *more* calories than you use up, you gain weight. If you take in *fewer* calories than you use, you lose weight by running a "calorie deficit." This is where the numbers come in.

This chapter will explain how to lose weight by increasing your calorie deficit. It will explain how to accurately track how many calories you take in by eating and how many you use up in a day. It will show you how to increase your weight loss by increasing exercise and also how to calculate how long weight loss will take.

There are a number of very good, sensible books on weight loss and nutrition on the market today. We will just look at the numbers part here. Before you embark on any program to lose weight, please consult your doctor.

Calorie Deficit—Your Key to Weight Loss

In theory, a calorie deficit is pretty easy to understand. Your daily calorie deficit is simply the number of calories you use up in a given day minus the number of calories you take in from eating that day.

Daily Calorie Deficit = calories used up minus calories taken in

If your weight is stable, your calorie deficit is zero. You are taking in about the same number of calories you are using up. If you want to lose weight, however, you have to create a calorie deficit. There are two ways to do that. Use up more calories by exercise or take in fewer calories by eating. It is up to you how you mix the two. For example, if you want to try to maintain a calorie deficit of 500 calories per day, you can eat 300 fewer calories and exercise away 200 more calories (300 plus 200 = 500). Or, if exercise just reminds you a little too much of junior-high gym class, you can maintain the same deficit just by eating 500 calories fewer each day and skipping the exercise. (However, it is important to note here that exercise has many other health benefits besides weight loss.)

Each pound of weight represents approximately 3,500 calories. So to lose ten pounds, you need to create a calorie deficit of 35,000 calories (10 times 3,500). Sound impossible? Take heart. The President's Council on Physical Fitness and Sports says that the average person takes in about 800,000 to 900,000 calories per year, so getting rid of 35,000 shouldn't be that hard. It's all about the numbers. Now, doesn't weight loss seem like a piece of cake? (Okay, bad analogy.)

Let's first go through the two elements of calorie deficit, calories taken in by eating and calories used up. Then we will run through some examples of how to use the calorie deficit approach to lose weight and how long it will take you to lose it.

Calories Taken In—Tracking Food Consumption

Studies now show that one of the most important aspects of successful weight loss is tracking food and calorie consumption. Simply put, you need to write down everything you eat each day, along with how many calories each food includes. We will explain how to do just that in the section "Keeping a Food Journal." Before we explain food journals, however, it is very important to get a handle on where the information that you put in the food journal is coming from. So, let's start by looking at food labels.

Reading Food Labels

Nutrition labels, while very helpful, can be terribly misleading and confusing. My husband and I spent forty-five minutes at the grocery store recently trying to figure out which popcorn was the most nutritious. The most important step in reading a food label is to determine how a "serving size" relates to how much you are actually eating. I was stunned to learn that a cup-sized soup on the market today, while giving the impression of being a single serving, is actually two servings according to its label. Since food labels include the number of calories per serving, you would need to double the calorie count if you ate the whole two-serving container.

Let's go through an example of how to use a food label to calculate the calories in a bowl of cereal.

PROBLEM

Today you are starting a new eating plan. (Again.) You are full of optimism and hope that this time it will really work. You have bought a food-tracking journal since you know how important it is to track your calorie intake during the day. Even though you are craving Belgian waffles, you carefully measure out a cup of low-fat, high-fiber Big Bran cereal into your bowl. You dutifully look at the food label on the box to see how many calories you will take in from the cereal. Help! It says the serving size is a ¾ cup, and you want to eat a cup. What do you do?

Let's first look at the nutrition label for Big Bran Cereal on page 25.

STEP 1

Determine the size of one serving of cereal from the first line of the food label.

¾ cup

STEP 2

Determine how much you are eating of the cereal.

1 cup

STEP 3

Find out how many servings of the cereal you actually ate by dividing the amount you ate (1 cup) by the food label serving size (¾ cup).

1 cup divided by ¾ cup = 1 divided by .75 = 1.33 servings

(Change ¾ into .75 so you can use your calculator.)

Nutrition Facts

Serving Size: 3/4 cup
Servings per Container: about 17

Amount per Serving	Cereal
Calories 90	
Calories from Fat 5	

	% Daily Value
Total Fat .5g	1%
Saturated Fat 0g	0%
Polyunsaturated Fat 0g	
Monounsaturated Fat 0g	
Cholesterol 0mg	0%
Sodium 230 mg	10%
Potassium 170mg	5%
Total Carbohydrate 23g	8%
Dietary Fiber 5g	20%
Soluble Fiber 0g	
Insoluble Fiber 5g	
Sugars 5g	
Other Carbohydrate 13g	
Protein 3g	

To change a fraction into a decimal just divide the number on the top (numerator) by the number on the bottom (denominator).

STEP 4

Get the nutritional information you are interested in from the appropriate line on the nutrition label.

Calories for one serving = 90

STEP 5

To get nutritional information for the food you actually ate, multiply the nutritional information on the food label from Step 4 (90) by the number of servings you ate from Step 3 (1.33).

90 calories times 1.33 = 120 calories

SOLUTION

You have taken in 120 calories.

This might seem like a lot of trouble to go through for each food. But, once you do it a few times, calculating nutritional information will seem second nature to you, and you will be able to do much of the process in your head.

In a Nutshell

READING FOOD LABELS

Multiplier = your actual serving size ÷ serving size from nutrition label on the food package

Calories in your serving = multiplier × calories in one serving of the food (from the nutrition label)

The same formulas can be used to calculate fat, sodium, and other nutritional information.

Labeling Claims—What Do They Mean?

The use of such terms as *low-fat, light (or lite), and percent fat-free* is highly regulated, but who knows what they really mean? The Food and Drug Administration (FDA) uses the following definitions:

Low-fat—3 grams or less of fat per serving.

Low-saturated fat—1 gram or less of saturated fat per serving.

Low-calorie—40 calories or fewer per serving.

Low-cholesterol—20 milligrams of cholesterol per serving *and* 2 grams or less of saturated fat per serving.

My favorite term that gets thrown around regularly is *percent fat-free*. We've all seen these claims: 98% fat-free. 83% fat-free. It sounds good, but what does it mean? According to the FDA, *percent fat-free* shows that the product is a low-fat or fat-free product and reflects the amount of fat grams in 100 grams of the food. In other words, if a food contains 3 grams of fat in a serving size of 100 grams, it contains 3 percent (3 divided by 100) fat. Therefore, it is 97 percent (100% minus 3%) fat-free.

To demonstrate how misleading labels can be I did some calculations on food in my cupboards and found one food that was 61-percent fat-free, based strictly on the percentage of fat grams in the weight of one serving. What was it? Regular-old, fat-laden potato chips.

Keeping a Food Journal

Now that you are able to calculate, based on the information on its label, the amount of calories you eat of a particular food, you can start your food journal. It will seem tedious at first, but once you get the hang of it, it won't be so bad. And remember, knowing how many calories are going in is an essential step to weight loss.

Let's walk through how to keep a food journal.

PROBLEM

Your twenty-fifth high school reunion is coming up in six months. You hear through the grapevine that Wanda Wiggleston, who stole your boyfriend Tommy from you when you were sophomores, is coming. You're over it, of course. Nonetheless, you need to look smashing for the event. You are reading this great new fitness book, *Terry's Tips for a Terrific Torso*, and Terry says the first tip to a terrific torso is to keep a food journal in which you write down everything you eat. Where should you start?

STEP 1

Create a four-column page for each day in your food journal (see page 29). In the first column, write down every item of food you eat and the amount you eat. (Be honest, now. If you eat the chocolate, it goes in the log.)

STEP 2

From a calorie guide or food label, record the calories per serving in the second column.

 You can find calorie calculators on the Internet for some food items not easily identified.

STEP 3

In the third column, calculate the number of servings you consumed by dividing the amount eaten by the amount in the serving.

1 cup of cereal divided by ½ cup in a serving = 2 servings of cereal

6 ounces of orange juice divided by 8 ounces in a serving = .75 servings

STEP 4

To find the number of calories for each food eaten, multiply calories per serving in column 2 by the number of servings in column 3. Record this number—the total calories in each food item eaten—in the fourth column.

150 calories per serving of cereal times 2 servings = 300 calories

120 calories per serving of orange juice times .75 serving = 90 calories

STEP 5

Total the number of calories for all foods eaten by adding up column 4.

The total number of calories taken in is 2,701.

■ SOLUTION

Your total calorie intake for one day is 2,701.

(Add the total calories of all 7 days to get your weekly total.)

Food Eaten	Calories/Serving	# of Servings	Total Calories
Cereal (1 cup)	150 for ½ cup	2	300
Skim milk (½ cup)	40 for ½ cup	1	40
Orange juice (6 oz.)	120 for 8 oz.	.75	90
Roast beef sandwich			
Bread (2 slices)	90 for 1 slice	2	180
Mayonnaise (1 tbs.)	100 for 1 tbs.	1	100
Lettuce (1 oz.)	4 for 1 oz.	1	4
Roast beef (5 oz.)	60 for 1 oz.	5	300
Tossed salad			
Lettuce (8 oz.)	4 for 1 oz.	8	32
Tomato (½)	20 for 1 tomato	½	10
Salad dressing (3 tbs.)	70 for 1 tbs.	3	210
Chocolate cookie (1 large)	100 for small cookie	4	400
Tortellini (8 pieces)	135 for 4 pieces	2	270
Red sauce (1 cup)	110 for ½ cup	2	220
Garlic bread (2 slices)	180 for 1 slice	2	360
Wine (1 glass)	95 for 1 glass	1	95
Popcorn (2 cups)	180 for 4 cups	.5	90
TOTAL			**2701**

Use your calculator as much as you can for these calculations. Sometimes it's easier to keep everything in fractions. It works either way. In the case of the orange juice, use your calculator to divide 6 by 8. In the case of the tomato, you guess that you ate about a half a tomato. One whole tomato has 20 calories, so ½ a tomato has 10. Many times you will have to estimate how much of a serving, or how many, you ate. Since your calorie chart only included small cookies, you guess your large cookie is equivalent to 4 small cookies and count it as 4 servings of cookies.

Use the same technique to track fat, sodium, or any other aspect of nutrition.

Calories Used Up

Now that you know how many calories you take in each day, let's turn our attention to how many calories you burn off each day. How many calories you use up in a given day is unique to each person and has a lot to do with weight, gender, muscularity, and activity level. Men tend to have higher calorie use because they are proportionately more muscular than women, which allows them to burn calories more efficiently. Using sophisticated equipment, a trained dietician can help you get a pretty good idea of your daily calorie use, and there are numerous Internet sites that will calculate this for you as well.

Calculating How Many Calories You Use Up in a Day

A rough way of calculating how many calories you use up in a day is to multiply your current weight by 10.

Calories Used Up in One Day = weight times 10

For example, if you currently weigh 165 pounds and do not exercise much, you probably use up about 1,650 calories each day (165 times 10).

If you are an active person, your multiplier might be closer to 12, and a very active person might have an even higher multiplier. As your weight-loss program continues, you can adjust the multiplier to be more accurate for you. I am a middle-aged woman who exercises most days, and my multiplier is 12.

In a Nutshell

CALORIES USED UP

Calories used up in a day = weight times 10

Increase multiplier if you are an active person.

Increasing Weight Loss by Exercise

There are a number of ways you can increase the calories you use up in a day by adding more activity to your life. By increasing the calories you use up in a day, you increase your calorie deficit and therefore increase your weight loss. Check out the chart below for an idea of the number of calories burned by participating in various activities. Remember, though: to lose weight you need to add *more* exercise. The exercise you are already doing is maintaining your current weight.

Use the figures in this exercise chart as a "general" idea of calorie use for each exercise type. The exact number of calories burned will vary greatly depending on gender, weight, fitness level, and intensity of exercise.

Let's use this chart in an example.

PROBLEM

You've just moved to a new city to start an exciting new job. You are feeling good about life and want to take this opportunity to lose some weight and develop some good habits around exercise. You'd like to start walking to and from work. You weigh about 135 pounds and are currently not physically active. It takes you about 45 minutes to walk each way. How many calories are you currently using up, and how many more will you use up on the days that you walk to and from work?

STEP 1

Estimate how many calories you are currently using up each day by multiplying your weight (135) by 10.

135 pounds times 10 = 1,350 calories per day

STEP 2

Calculate the numbers of hours you will walk on a given day, adding together the time it took to walk to and from work.

Calories Expended by Exercise Type

Exercise Type	Calories per hour*
A. Sedentary Activities	
Lying down or sleeping	90
Sitting quietly	84
Sitting and writing, card playing, etc.	114
B. Moderate Activities	**(150–350)**
Bicycling (5 mph)	174
Canoeing (2.5 mph)	174
Dancing (ballroom)	210
Golf (two-some, carrying clubs)	324
Horseback riding (sitting to trot)	246
Light housework, cleaning, etc.	246
Swimming (crawl, 20 yards/min)	288
Tennis (recreational doubles)	312
Volleyball (recreational)	264
Walking (2 mph)	198
C. Vigorous Activities	**More than 350**
Aerobic dancing	546
Basketball (recreational)	450
Bicycling (13 mph)	612
Circuit weight training	756
Cross-country skiing (5 mph)	690
Football (touch, vigorous)	498
Ice-skating (9 mph)	384
Jogging (10 minute mile, 6 mph)	654
Racquetball	588
Roller-skating (9 mph)	384
Scrubbing floors	440
Swimming (crawl, 45 yards/min)	522
Tennis (recreational singles)	450

*Hourly estimates based on calculations for calories burned per minute by a 150-pound (68kg) person.

Sources: William D. McArdle, Frank I. Katch, and Victor L. Katch, *Exercise Physiology: Energy, Nutrition and Human Performance* (2nd edition) (Philadelphia: Lea & Febiger, 1986); Melvin H. Williams, *Nutrition for Fitness and Sport* (Dubuque: William C. Brown Company Publishers, 1983).

45 minutes plus 45 minutes = 90 minutes = 1.5 hours

(To convert minutes to hours, divide the number of minutes by 60, since there are 60 minutes in an hour.)

STEP 3

Find the calories expended per hour of walking on the chart above.

198 calories per hour

STEP 4

To find the total calories you will use up by exercising on a given day, multiply the hours you calculated in Step 2 (1.5) by the number of calories burned per hour from Step 3 (198).

1.5 hour times 198 calories = 297 calories

STEP 5

To calculate the total number of calories you will use up on the days you walk to and from work, add the calories you are currently using up without exercise from Step 1 (1,350) to the extra calories you will use up from exercise from Step 4 (297).

1,350 calories plus 297 calories = 1,647 calories

SOLUTION

You will use up 1,647 calories on days you walk to and from work.

(Way to go.)

You can increase your calorie deficit by increasing the intensity of your exercise. In the Calories Expended by Exercise Type chart, you can see that if you bicycle at 5 miles per hour, you burn 174 calories per hour, but if you bicycle at 13 miles per hour, you can burn 612 calories.

Exercising "In the Zone"

Often, fitness experts will recommend that you exercise "in the zone," and I don't think they mean the Twilight Zone. As far as I can tell, it means to keep your exercise at a level of intensity where you can hardly breathe and feel like you are about to lose your lunch. If you would like to take a more scientific approach, exercising in the zone usually means keeping your pulse rate at a certain percentage of your maximum heart rate (or the fastest rate at which your heart can beat). Exercising in the zone is a great way to increase the intensity of your exercise and therefore increase the number of calories you use up in a day.

According to the President's Council on Physical Fitness and Sports, beginners should start at a level of 60 percent of their maximum heart rate, and the more advanced can exercise at 80 percent. As always, consult your doctor before you start any exercise program.

Maximum Heart Rate

You can purchase sophisticated equipment that will give you an accurate measurement of your maximum heart rate. An easy way to get an approximate figure is to subtract your age from 220.

Maximum Heart Rate = 220 minus age

(Of course, to do this right you will need to be honest about your age.)

"The Zone"

To calculate where to keep your pulse rate in order to stay in your exercise "zone," multiply the desired heart rate percentage by your maximum heart rate.

Desired Pulse Rate = desired percentage times
maximum heart rate.

PROBLEM

You are 46 years old and have just gotten your pictures
back from your second honeymoon in Jamaica. You can't
believe that the guy in the picture with the belly obscuring
his belt is really you. When did this happen? You vow to
hit the gym and make fast progress. The sign at the gym
says that the key to fitness success is to exercise intensely
at 60% to 80% of your maximum heart rate. Your doctor
gives you clearance to start a vigorous exercise program.
Where should you keep your pulse rate?

STEP 1

Calculate your maximum heart rate by subtracting your age
(46) from 220.

220 minus 46 = 174

STEP 2

Multiply your desired minimum and maximum intensity
(60% and 80%) by your maximum heart rate.

60% times 174 = .60 times 174 = 104

80% times 174 = .80 times 174 = 139

(For ease in multiplying, convert percentages to decimals by
dividing the percentage by 100.)

 A "percent" is just a number divided by 100. Use your
calculator, and it's actually pretty easy.

SOLUTION

You should keep your pulse rate between 104 and 139
per minute depending on how fit you are and what your
exercise goals are.

In a Nutshell

EXERCISING IN THE ZONE

Maximum heart rate = 220 − age

Desired pulse rate = desired heart rate percentage × maximum heart rate

---

Using Calorie Deficit to Lose Weight

Now that you know how many calories you take in by eating and how many you use up in a given day, you can calculate your daily calorie deficit and use it as an important tool for weight loss. It is really very easy. Remember the formula from earlier:

Daily Calorie Deficit = total calories used up in one day minus calories taken in in one day

If your weight is currently stable, your calorie deficit should be around zero. That means the calories you are taking in by eating are about equal to the amount of calories you use up in a day. For example, if you take in 2,000 calories from food in a day, and use up about 2,000 calories, your daily calorie deficit is zero (2,000 minus 2,000 = 0).

If you want to lose weight, you need to create a calorie deficit by either taking in fewer calories from eating or using up more calories. Let's say you have followed the instructions above to figure out that you currently use up about 1,850 calories a day. You think you can use up another 150 calories a day by increasing your exercise. You will then use up a total of 2,000 calories (1,850 plus 150) on an average day. You think you can get your food consumption down to about 1,500 calories. Your daily calorie deficit will then be 500 calories (2,000 minus 1,500). To calculate your weekly calorie deficit, just multiply your average daily calorie deficit by 7.

Weekly Calorie Deficit = daily calorie deficit times 7

If your daily calorie deficit is 500 calories, then your weekly calorie deficit will be 3,500 (500 times 7).

PROBLEM

You are a big guy and have always been told that "you carry your weight well," but now it is beginning to look like you are "carrying" a little too much. Your weight is currently 245 pounds, and you would like to get down to 195 by adopting a new fitness plan of exercising and eating well. You are fairly active but want to start running a half hour each day at a moderate pace to increase your weight loss. You have decided that you could safely maintain a 2,000-calorie-a-day diet. What will your daily and weekly calorie deficits be?

STEP 1

Determine the amount of calories you are currently using up by multiplying your weight (245) by the appropriate multiplier. (See page 30.) **You decide on a multiplier of 11 because you are fairly active.**

11 times 245 pounds = 2,695 calories

STEP 2

Determine how many additional calories you would use up if you started running moderately for a half hour each day by looking on the Calories Expended by Exercise Type chart on page 32.

One hour of jogging uses up 654 calories, so ½ hour would use up ½ of 654, or 327, calories.

STEP 3

Add the results of Step 1 (2,695) and Step 2 (327) to determine how many total calories you would be using up in your new fitness program.

2,695 calories plus 327 calories = 3,022 calories

STEP 4

To calculate your *daily* calorie deficit under your new plan, subtract the number of calories you expect to take in from eating (2,000) from the total number of calories you expect to use up from Step 3 (3,022).

3,022 calories minus 2,000 calories = 1,022 calories

STEP 5

To calculate your expected *weekly* calorie deficit, multiply your expected *daily* calorie deficit from Step 4 (1,022) by 7.

1,022 times 7 = 7,154 calories

SOLUTION

Your daily calorie deficit will be 1,022 calories under your new fitness plan.

Your weekly calorie deficit will be 7,154 calories under your new fitness plan.

Keeping up a steady weight loss requires you to consistently maintain your desired weekly calorie deficit. You can do that by carefully tracking the amount of calories you take in and use up each day, using the techniques described above. Unfortunately, the numbers do not lie. If you eat 300 calories over your planned intake one day, you need to make up for that by eating 300 less than planned intake another day or by exercising an additional 300 calories away.

Keeping track of your calorie deficit is a great way of making you aware of all the elements of healthy weight loss. These formulas are not cast in stone, however, and you should be patient with yourself as you work through the weight-loss process. Maintaining even a small weekly calorie deficit will eventually yield big results.

In a Nutshell

CALCULATING CALORIE DEFICIT
Daily calorie deficit = total calories used up in one day – calories taken in from eating in one day

Weekly calorie deficit = daily calorie deficit × 7

How Long Will It Take You to Lose Weight?

Now that you know your weekly calorie deficit, it is easy to figure out about how long it will take you to lose that weight you would like to shed. A pound of weight represents about 3,500 calories. To figure out how many pounds you can lose in an average week, just divide the weekly calorie deficit you plan to maintain by 3,500.

Pounds Lost in One Week = calorie deficit for one week divided by 3,500

If you maintain a calorie deficit of 3,500 calories a week, you will lose about one pound in a week (3,500 divided by 3,500). If you go hog-wild and maintain a calorie deficit of 7,000 calories a week, then you can lose two pounds in a week (7,000 divided by 3,500). If you want to take it slowly and maintain a calorie deficit of 1,750 per week, you will lose a half pound a week (1,750 divided by 3,500 = .5, or ½).

Be careful not to lose weight too quickly. An abrupt weight loss might not be healthy and is probably harder to sustain. Weight loss experts generally recommend a one- to two-pound weight loss per week.

Once you know your average weight loss per week, it is easy to figure out how long it will take to lose a certain amount of weight. To figure out how many weeks it will take you to lose the weight you want, just divide your weight loss goal by the number of pounds you expect to lose in one week.

Number of Weeks to Lose Desired Weight = weight loss goal in pounds divided by the number of pounds lost in one week

If you lose a pound a week and want to lose twenty pounds, it will take you about twenty weeks (20 divided by 1) to do that. If you lose two pounds a week, it will take you only ten weeks to lose the twenty pounds (20 divided by 2).

PROBLEM

Your fortieth birthday is coming up. People have started telling you that you look great "for your age." You weigh about 150 pounds, and you would love to lose 15 pounds. You decide that to lose the weight you will maintain a daily calorie deficit of about 400 calories through increased exercise and reduced eating. How long will it take you to lose the 15 pounds?

STEP 1

To calculate your *weekly* calorie deficit, multiply your *daily* calorie deficit (400) by 7.

400 calories times 7 = 2,800 calories per week

STEP 2

Calculate the number of pounds you will lose in a week by dividing your weekly calorie deficit from Step 1 (2,800) by 3,500.

2,800 calories divided by 3,500 calories = .8 pounds per week

STEP 3

Calculate the total number of weeks to lose the entire weight by dividing the total number of pounds you want to lose (15) by the number of pounds you will lose in one week from Step 2 (.8).

15 pounds divided by .8 pounds per week = 18.75 weeks Round up to 19 weeks.

STEP 4

Convert weeks to months by dividing the number of weeks of weight loss from Step 3 (19) by an average of 4 weeks in a month.

19 weeks divided by 4 = 4.75 months

SOLUTION

It will take you between 4 and 5 months to lose 15 pounds by maintaining a daily calorie deficit of 400.

In a Nutshell

HOW LONG WILL IT TAKE YOU TO LOSE WEIGHT?

Pounds lost in one week = weekly calorie deficit ÷ 3,500

Number of weeks to lose desired weight = weight-loss goal in pounds ÷ number of pounds lost in one week

---

Weight loss is tough going, regardless of the plan you follow. Understanding the relationship between the number of calories taken in and the number used up should, at the very least, make you more aware of how weight creeps up on us (usually from behind). Knowing the numbers might also makes losing that weight a little easier for you.

3 Home Improvement

You've seen the pictures in the home improvement magazines—happy couples lounging in their beautiful, newly refurbished living rooms. This chapter cannot guarantee that kind of bliss, but it will get you started, teaching you how to determine the quantity of materials needed for various projects and how to figure out the total cost.

Although home projects differ, the materials needed generally break down into three types: materials that are measured in lengths (like wire, fringe, molding, fencing, and lumber); materials that come in sheets (like carpeting, flooring, fabric, and wallpaper); and materials that you buy in bulk (like mulch, stone, sand, and gravel). All you have to do is figure out whether your material is sold in lengths, sheets, or in bulk, and go to the appropriate section within this chapter.

This chapter will also help you buy paint, as well as help you space things at even intervals, like pictures on a wall, racks in a closet, flowers in a garden, or children in a family. (Just kidding about the children.) You will also learn how to easily change from one kind of measurement to another.

Changing Units of Measure

The first thing you need to know about home improvement projects is that you will often need to change between different units of measure. You might need to go from inches to feet, from feet to yards, or even deal with more complicated conversions like cubic feet to cubic yards (ick). Don't panic though. Just use the simple Home Improvement Changing Table below, and you'll never have to worry about it again.

Changing Table

HOME IMPROVEMENT

To go from ...	To ...	You ...	.....	By ...
inches	feet	divide	inches	12
inches	yards	divide	inches	36
feet	inches	multiply	feet	12
feet	yards	divide	feet	3
yards	inches	multiply	yards	36
yards	feet	multiply	yards	3
square inches	square feet	divide	square inches	144
square feet	square inches	multiply	square feet	144
square feet	square yards	divide	square feet	9
square yards	square feet	multiply	square yards	9
cubic feet	cubic yards	divide	cubic feet	27
cubic yards	cubic feet	multiply	cubic yards	27

PROBLEM

You are all set to put a beautiful new floor in your kitchen. You have measured your room carefully and calculated that you need 120 square feet of flooring. So far, so good. But when you get to the store, the flooring you like is sold in square yards. How many square yards of flooring do you need?

■ STEP 1

Find the line in the Home Improvement Changing Table above for changing square feet to square yards.

To go from . . .	to . . .	you . . .	. . . . .	by . . .
square feet	square yards	divide	square feet	9

■ STEP 2

Following the instructions on the chart, divide 120 square feet by 9.

120 square feet divided by 9 = 13.3 square yards

Round up to 14.

■ SOLUTION

Buy 14 square yards of flooring.

Certain types of flooring cannot be bought based strictly on area, as in the example above. Many types, such as linoleum and carpeting, come in specific widths. See the section below called "Things That Come in Widths" to compute flooring that comes in specific widths.

How Much Do You Need and How Much Will It Cost?

Now that you are comfortable with working with different kinds of measurements, you can begin to figure out how much you need of various materials and how much the materials will cost. Let's start by calculating how much you need of materials that are measured in length.

Things That Come in Lengths (Wire, Fringe, Molding, Fencing, and Lumber)

Calculating how much you need of materials that are measured by length is pretty straightforward unless the material you need comes in special sizes. Fringe, wire, rope, string, telephone line, and the like are available in any length. To determine how much you need of one of these, simply calculate the distance it will need to run or the perimeter of (that's the distance around) the object you're working with. Other materials, like molding and certain types of fencing and lumber, are often available only in premeasured lengths, so calculations become more complicated. Below, you'll see examples of how to do both.

Materials You Can Buy in Exact Measurements

PROBLEM

You have just bought a beautiful new couch, and you and your teenaged daughter go to the store to pick up a few throw pillows to go with it. Astounded by the price of the pillows, you boldly state that you could make pillows for half of what it would cost to buy them. Never missing an opportunity to call her mom on something, your daughter says, "So, why don't you, Mom?" Your pride having taken over, you march over to the fabric store, pick out some fabric, and find some beautiful fringe to go with it. The fringe costs $5 per yard. How much fringe do you need to decorate the three pillows, and how much will it cost? One pillow is square, with each side measuring 18 inches. Another is rectangular and is 14 inches by 10 inches. The third is a circle that measures 12 inches across.

To calculate the perimeter of a *square*, multiply the length of one side by 4. To calculate the perimeter of a *rectangle*, multiply the long side by 2, multiply the short side by 2, and then add the two numbers together. To calculate the perimeter of a *circle*, multiply the diameter (the length from one side to the other) by 3.14 (this number is called *pi*).

STEP 1

Calculate the perimeter (length around) of the square pillow by multiplying the length of one of the sides (18) by 4.

Perimeter of square pillow = 4 times 18 inches = 72 inches

STEP 2

To calculate the perimeter of the rectangular pillow, first multiply the length of the long side (14) by 2, and multiply the length of the short side (10) by 2. Then add those two numbers together.

14 inches times 2 = 28 inches

10 inches times 2 = 20 inches

perimeter of rectangular pillow = 28 inches plus 20 inches = 48 inches

STEP 3

Calculate the perimeter of the circular pillow by multiplying the diameter (12) by 3.14.

perimeter of circular pillow = 12 inches times 3.14 = 37.68 inches

Round 37.68 inches up to 38 inches.

STEP 4

To get the total amount of fringe needed, add the perimeters of all three pillows.

72 inches plus 48 inches plus 38 inches = 158 inches

STEP 5

Since the cost of the fringe is in yards, and the perimeters you calculated are in inches, consult the Home Improvement

Changing Table on page 44 to convert inches to yards. Following the table, divide the total inches from Step 4 (158) by 36 to get total yards.

158 inches divided by 36 = 4.4 yards

Round up to 5 yards.

STEP 6

To calculate your cost for the fringe, multiply the number of yards needed from Step 5 (5) by the cost per yard ($5).

5 yards times $5 per yard = $25

SOLUTION

You need 5 yards of fringe, and it will cost $25.

Materials That Come in Premeasured Lengths
Now that you have mastered buying materials that are sold by length, let's look at materials that are sold in premeasured lengths, like molding, lumber, certain framing materials, and the like.

PROBLEM

Your wife wants to have some crown molding put up in the living room and has started calling around for estimates. In a moment of bravado, you proclaim, "No need to call someone in, dear. I can do that." Your wife is skeptical, particularly after her mother had that unfortunate incident with the toilet you installed in her house. But she's willing to give you a chance to redeem yourself. You measure the room, and find that its length is 22 feet and its width 16 feet. When you get to Howie's Home Store, however, you find that the molding you like is sold in 12-foot lengths, which cost $2 a foot. How much molding do you need, and how much will it cost?

To solve this you will need to figure out how to maximize the use of full-length boards while minimizing the number of places where the molding meets to form a seam. It helps to sketch your room using graph paper.

STEP 1

Draw a picture of the room.

STEP 2

Start by seeing how many full-length boards will fit on each wall.

The two 22-foot walls can take one 12-foot length each, making two total.

The two 16-foot walls can take one 12-foot length each, making two total.

Number of 12-foot boards needed for this step: 4.

STEP 3

Calculate the length of the parts of the walls not covered by the full 12-foot lengths by subtracting the 12-foot lengths from the total wall measurement.

The two 22-foot walls each have 10 feet left (22 minus 12 = 10).

The two 16-foot walls each have 4 feet each left (16 minus 12 = 4).

STEP 4

Looking at your picture, figure out how to cover the remaining space on each wall with the least number of seams.

Each of the two 22-foot walls needs one board to cover the remaining 10 feet (2 times 1 board = 2 boards). Each of the two 16-foot walls have 4 feet left (2 times 4 feet = 8 feet), so both can be covered with one additional 12-foot board.

Number of boards needed for this step: 3.

STEP 5

Calculate the total number of boards you need by adding the number of boards from Step 2 (4) and the number of boards from Step 4 (3).

4 boards plus 3 boards = 7 boards

STEP 6

To figure out the cost of the molding, first calculate the total cost of one board by multiplying cost per foot ($2) times number of feet in one board (12).

$2 times 12 = $24 per board

STEP 7

Calculate the total cost of the boards by multiplying the cost for one board from Step 6 ($24) by the number of boards needed from Step 5 (7).

$24 times 7 boards = $168

SOLUTION

You need 7 12-foot boards, which will cost $168.

Allow extra length for molding and other materials that need to be cut on an angle (where corners meet). Give yourself at least an extra 6 to 12 inches on each wall. In this example, there is plenty left over to cover any extra needed.

In a Nutshell

THINGS THAT COME IN LENGTHS

1. Find the length of the material needed by measuring the distance of the path along which the material will go.

2. If the path is the distance around (the perimeter of) an area, use one of the following formulas:

   Perimeter of square = 4 × width

   Perimeter of rectangular = (2 × length) + (2 × width)

   Perimeter of circle = 3.14 × diameter (measurement across a circle)

   Perimeter of an odd shape = add measurements of each side

3. If materials needed come in premeasured lengths, calculate how many whole premeasured lengths you need per side and then how many lengths you need to cover the rest. It helps to draw the area on graph paper.

4. To calculate cost, multiply cost per unit times total units needed. Use the Home Improvement Changing Table on page 11 to convert different units of measure.

When figuring out how much material you need, always make sure you add extra for imperfections in the materials or for mistakes. Home improvement projects, unfortunately, lend themselves to do-overs. Make sure you have enough.

Things That Come in Sheets (Carpeting, Flooring, Curtains and Drapes, and Wallpaper)

Now that you know how to buy materials that cover a path, let's move on to buying materials that cover a specific area, like a floor or a window or the like. Most of the areas you will want to cover are in the shape of a rectangle. You might remember that you calculate the area of a rectangle by multiplying its width by its length. It would be nice, then, if you could just calculate the area of the rectangular surface using this simple formula and be done with it. Unfortunately, most things that cover surfaces, like carpeting, flooring, wallpaper, and fabric, come in either premeasured widths or sheets, so the calculations can get kind of complicated.

Let's start with a common area cover—carpeting. It most often comes in 12- or 15-foot widths, called "runs." The first example below walks you through a situation in which your room is not quite as wide as one run. The second example takes you through the more common situation where your room is wide enough to require more than one run.

Your Room Width Is Less than or Equal to One Width (Run) of Carpeting

PROBLEM

You decide it's well past time to replace the carpeting in the den. It's getting kind of worn, and there's that red stain from Junior's juice box. (Junior is now married with kids of his own.) You go to Candy's Custom Carpeting and pick out a lovely carpet, which sells for $25 per square yard and comes in 12-foot runs. How much carpeting will you need, and about how much will it cost?

STEP 1

Measure the length and width of the room at their longest points in feet. Round up to the next foot to allow room for error.

The room is 11 feet wide.

The room is 13 feet, 7 inches long. Round up to 14 feet.

 It is often very helpful to draw a picture of your room as you make these calculations.

STEP 2

Determine how many runs of carpet you need.

The room width (11 feet) is less than the width of the run (12 feet).

Therefore, you need 1 run.

STEP 3

Determine the square feet of carpeting you need by multiplying the width of the carpet run (12 feet) by the length of your room (14 feet).

12 feet times 14 feet = 168 square feet

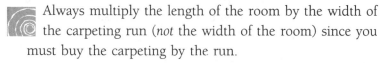 Always multiply the length of the room by the width of the carpeting run (*not* the width of the room) since you must buy the carpeting by the run.

STEP 4

You now know how many square feet of carpeting you need, but Candy sells carpet by the square yard. Using the Home Improvement Changing Table on page 44, change square feet into square yards by dividing the number of square feet from Step 3 (168) by 9.

168 square feet divided by 9 = 18.7 square yards

Round up to 19 square yards.

STEP 5

Determine the total cost of the carpet by multiplying the number of square yards from Step 4 (19) by the cost per yard.

19 square yards times $25 per yard = $475

SOLUTION

You will need 19 square yards of carpet, and it will cost $475.

Your Room Width Is Greater Than One Width (Run) of Carpeting

PROBLEM

You have the carpet installed in the den, and you are so satisfied with the job that you did you are thinking about installing the same $25-per-square-yard carpeting, which comes in 12 feet runs, in your living room. How much will you need, and what will it cost?

STEP 1

Measure the length and width of your room at their longest points. Round up to the next foot to allow for error.

The room is 14 feet wide.

The room is 19 feet, 3 inches long.

Round up to 20 feet long.

STEP 2

To minimize cost, you want to determine how to lay out the carpet to use the least amount of carpeting. To do this, calculate both how much carpeting you would need if you lay it from the short side and how much if you lay from the long side.

(Drawing a picture of your room will help you visualize this.)

1. As you can see from Option 1, rolling out the carpet from the short (14 feet) side requires two 12-foot runs, each 20

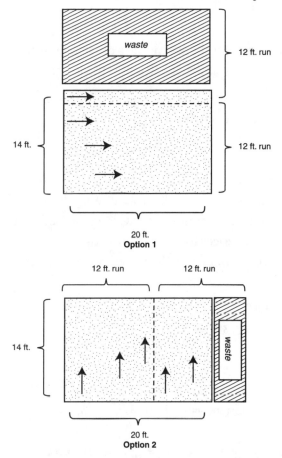

feet in length. Calculate the total carpeting needed for this option by multiplying the width of the carpeting needed (2 runs = 2 times 12 feet = 24 feet) by the length (20 feet).

24 feet times 20 feet = 480 square feet

2. Looking at Option 2, rolling out the carpet from the long (20 feet) side requires two 12-foot runs, each 14 feet in length. Calculate the total carpeting needed for this option by multiplying the width of the carpeting needed (2 runs = 2 times 12 feet = 24 feet) by the length (14 feet).

24 feet times 14 feet = 336 square feet

STEP 3

Consider the two alternative ways of laying out the carpet from Step 2, and choose the one that uses the lesser amount of carpet.

Option 2 uses the lesser amount of carpet, at 336 square feet. Lay the carpet out from the long end.

STEP 4

You now know how many *square feet* of carpeting you need, but remember, Candy sells her carpeting by the *square yard.* Using the Home Improvement Changing Table on page 44 to convert square feet to square yards, divide the number of square feet from Step 3 (336) by 9 to get square yards needed.

336 square feet divided by 9 = 37.3 square yards

Round up to 38 square yards.

STEP 5

Determine the total cost of the carpet by multiplying the number of square yards from Step 4 (38) by the cost per yard ($25).

38 square yards times $25 per yard = $950

SOLUTION

You will need 38 square yards of carpeting, and it will cost $950.

You can see how expensive carpeting can get when your room width is greater than the width the carpeting comes in. And, there is a lot of waste. Talk to the carpet salesperson about remnants that might be available. You might not have to buy an entire extra roll. You can also use smaller remnants for oddly shaped areas.

Flooring

Flooring, such as vinyl flooring, is sold similarly to carpeting and normally comes in 6-foot or 12-foot widths. Check to see what width your flooring comes in, and follow the step-by-step examples for carpeting.

Curtains and Drapes

Calculating the amount of fabric you need for curtains and drapes can be tricky. It is best to consult a salesperson at the fabric store or a book on window coverings to determine how to measure your window and how to allow for pleats, gathers, hems, and headings. A rule of thumb for gathers or pleats is to double the width of your window; for hems, allow three inches; for headings it depends on the treatment you want to use.

Fabric comes in standard widths, similar to the standard carpeting runs discussed earlier in this chapter. The area of the window is similar to the area of the floor covered by the carpeting. Once you know the width and length of the fabric you need, you can use the carpeting examples above to help you calculate how much you need and to determine how much it will cost.

Let's say you want to calculate the amount of fabric you need for pleated drapes over a window that is 60 inches wide. You would first double the width to 120 inches to allow for the pleats. It would be easy if you chose a fabric that came in 60-inch width. You would just buy two widths (2 times 60 = 120 inches) in the length you needed and you would be all set. However, if you chose fabric that comes in the more common width of 48 inches, two widths (2 times 48 = 96 inches) would not be enough. You would need three widths (3 times 48 = 144 inches) in order to provide the 120 inches of fabric necessary for the fullness you want.

To calculate the length you need, measure from the top to the bottom of the area you want the drape to cover. Then add enough length to make the top pocket and the bottom hem. In addition, if your fabric has a repeat pattern that is equal to or longer than twelve inches, you will need to add 25 percent more length by multiplying the total length of fabric needed by 1.25. (Check books on making drapes and curtains to determine the exact allowance you need for hems, headings, and patterns.) And finally, remember to add a bit for error.

Wallpaper

Wallpaper is different from carpeting, flooring, and fabric in two ways. First, there is less waste because you can wrap it around corners. Second, wallpaper is sold in narrow "rolls" which are usually measured by the square foot. Therefore, you will be less concerned with the width of the roll and more concerned with how many square feet are in a roll of paper. Basically, the trick is to figure out how many square feet you will need to cover an area and then how many rolls that will take.

Many manufacturers package their wallpaper in double, or even triple, rolls, thereby causing you to buy more than you may need. Be sure to ask exactly how the paper you want is sold.

PROBLEM

The kids are at summer camp, so you and your husband decide it would be a great time to do that wallpapering job in the dining room that you've been putting off. (Sounds to me like it would be a great time to go on a cruise, instead of wallpapering your dining room. But hey, that's your business.) You've finally found some paper you can both agree on (no small feat) and you need to know how much to buy in order to have plenty for mistakes. A single roll includes 30 square feet of paper, and it costs $20 per roll. How much do you need, and how much will it cost?

STEP 1

Measure the height, length, and width of your room, rounding up to the next foot if the measurement is not in whole feet. Do not subtract doors, windows, and other openings.

The height of your room is 8 feet.

The length of your room is 17 feet.

The width of your room is 15 feet.

STEP 2

Calculate the perimeter of the room (the distance around the room) by adding the two lengths (17 feet) and two widths (15 feet).

perimeter = (2 times 17 feet) plus (2 times 15 feet)

perimeter = 34 feet plus 30 feet = 64 feet.

To calculate the perimeter of (the distance around) a rectangle, multiply the long side by 2, then multiply the short side by 2, and add the two numbers together.

STEP 3

Calculate the total area of the walls by multiplying the perimeter of the room from Step 2 (64) by the height of the room from Step 1(8).

64 feet times 8 feet = 512 square feet

STEP 4

Calculate the number of rolls of wallpaper needed by dividing the total area in square feet from Step 3 (512) by the amount of square feet in one roll (30).

512 divided by 30 = 17.1 rolls

To give yourself plenty of wallpaper, round up to 18 rolls.

STEP 5

Compute the cost of the wallpaper by multiplying the number of rolls from Step 4 (18) by the price of one roll ($20).

18 times $20 = $360

SOLUTION

You need 18 rolls of wallpaper, and it will cost $360.

You might need to add more paper if the paper has a pattern. Check the package for the pattern's repeat size. If the pattern's repeat size is larger than 12 inches, add 25% more wallpaper by multiplying the number of rolls needed by 1.25.

In a Nutshell

THINGS THAT COME IN SHEETS

Sheets that come in specific widths: carpeting, flooring, fabric

1. Measure the width and length of the surface you are covering.

2. Draw a picture of the surface you are covering.

3. Determine how many sheets (widths) of the material you need by seeing how the width of the surface compares to the width of the material. If you have the option of laying the material either widthwise or lengthwise, calculate how many sheets (widths) you'd need for each option.

4. Calculate the amount of material needed.

   Amount of material needed = number of sheets × width of one sheet × length of the surface to be covered.

continued

---

continued from previous page

If you have the option of laying the material either widthwise or lengthwise, calculate the amount of material needed for each option so you can choose the one that uses less material.

5. Cost = amount of material needed × cost for one unit of the material needed.

Sheets that come in rolls: wallpaper

1. Calculate the total area of the walls to be papered.
    Area of walls (square feet) = perimeter (the distance around) in feet × height in feet

2. Calculate number of rolls needed.
    Number of rolls needed = area of walls in square feet ÷ square feet in one roll

3. Total cost = number of rolls × the cost of one roll.

Make sure you measure the area in the same units (square feet, square yards) as the material is sold. If not, use the Home Improvement Changing Table on page 44 to convert your measurements.

Things That Come in Bulk (Mulch, Gravel, Stone, and Sand)

Now that you have figured out how to work with materials that come in lengths and sheets, let's turn to the last kind of materials— things that come in bulk. Things that come in bulk are most often measured in cubic yards. While sometimes palms start to sweat at the very notion of dealing with "cubic yards," it's really not so bad. All you have to do is break the calculation into smaller parts. There are really just two things you need to know: the measurements of the space you are trying to cover and how high you want to pile the material you are using.

PROBLEM

Spring is in the air. The birds are chirping. Trees are budding. Flowers are blooming. People are happily working in their yards. But where are you? Frantically scratching numbers on a piece of paper, trying to figure out how much mulch you need for your garden. The neighbors are still chuckling about that load of mulch you ordered last year that took up your entire driveway. You want to avoid a similar disaster this year, but where do you start? Monty's Mulch charges $20 per cubic yard. How much do you need, and how much will it cost?

STEP 1

Measure the length and width of the rectangular garden beds you want to mulch.

Bed A is 20 feet long and 5 feet wide.

Bed B and Bed C are 12 feet long and 6 feet wide.

Bed D is 30 feet long and 6 feet wide.

STEP 2

Calculate the area of each bed by multiplying the length of each by its width, from Step 1.

Area of Bed A = 20 feet times 5 feet = 100 square feet

Area of Beds B and C = 12 feet times 6 feet = 72 square feet each

Area of Bed D = 30 feet times 6 feet = 180 square feet

STEP 3

Calculate the total area by adding the areas of all the beds from Step 2 together.

100 square feet plus 72 square feet plus 72 square feet plus 180 square feet = 424 square feet

(You add 72 twice since there are two beds that size.)

STEP 4

Now that you know the area you are covering, determine how thick you want the mulch.

You decide you want it to be about 3 inches thick.

STEP 5

You now have one measurement in feet and another in inches. This is very common. It will be easier to work with feet than inches. Following the Home Improvement Changing Table on page 44, divide the thickness in inches from Step 4 (3) by 12 to convert from inches to feet.

3 inches divided by 12 = .25 feet

Always make sure you have all your measurements in the same units. It does not really matter which unit you choose, as long as all the measurements are in the same one. Use the Home Improvement Changing Table on page 44 if you need some help converting from one unit to another.

STEP 6

Now that you have all the measurements in the same units, you can calculate the volume of mulch you need. To calculate the volume of mulch needed, multiply the area to be covered from Step 3 (424 square feet) by the thickness of coverage from Step 5 (.25 feet).

424 square feet times .25 feet = 106 cubic feet

 Don't let the word "cubic" make you nervous. It's just a measurement for volume.

STEP 7

Since Monty sells his mulch in cubic yards, not cubic feet, you need to convert the volume from cubic feet into cubic yards. Following the Home Improvement Changing Table on page 44 to convert cubic feet to cubic yards, divide cubic feet from Step 6 (106) by 27.

106 cubic feet divided by 27 = 3.9 cubic yards

Round up to the next whole yard and order 4 cubic yards.

STEP 8

Now that the volume and price are both in the same units (cubic yards), you can figure out how much the mulch will cost. To determine the cost of the mulch, multiply the volume you need in cubic yards from Step 7 (4) by Monty's price for one cubic yard ($20).

4 cubic yards times $20 per cubic yard = $80

SOLUTION

You will need 4 cubic yards, and it will cost you $80.

For small jobs, mulch is also sold in bags that contain 3 cubic feet. If this is how you're buying it, stop at the end of Step 6. To see how many bags you would need, divide the total cubic feet you need by 3, since there are 3 cubic feet in each bag. To calculate the cost, multiply the number of bags you need by the cost of one bag.

Often the shapes of flower beds, patios, and driveways are not easily defined, making it difficult to calculate their area. In that case, measure one dimension of the area, then draw the shape on a piece of graph paper as close to scale as you can. You might let every square of graph paper represent one foot. Then, simply count the number of squares the area covers on the graph paper. This number is the approximate area in square feet. Just take this number and start from Step 4 in the process above to find out how much mulch you need to cover the area.

In a Nutshell

THINGS THAT COME IN BULK

1. Measure the surface you want to cover (in feet).

2. Calculate area to be covered.

    Area of square = width × width

    Area of rectangle = width × length

    Area of circle = 3.14 × radius × radius (radius = length from center of circle to the edge)

3. Determine how thickly you want to cover the area (in inches).

4. Convert thickness to feet.

    Thickness (in feet) = thickness in inches ÷ 12

5. Calculate volume of material needed.

    Volume of material (in cubic feet) = area from #2 × thickness from #4

6. If sold in cubic yards, convert volume from #5 to cubic yards by dividing by 27.

7. Calculate cost.

    Cost = volume of material × cost per cubic unit (feet or yards)

Paint

Calculating the amount of paint you need is not much different from calculating other materials. You really just need to know two things: the area of the surface you are covering and how much one gallon of paint can cover.

PROBLEM

Your brother and his new wife are visiting your home for the first time, and you want to make the visit special. You have wanted to paint the guest room, and their visit offers a great incentive. You measure the room and find it is 16 feet wide, 22 feet long, and 10 feet tall. You pick out the color and read on the back of the can that a gallon of paint should cover 400 square feet. Each gallon can costs $25. How much paint will you need for two coats of coverage, and how much will it cost?

STEP 1

Calculate the perimeter of the room (the distance around the room) by adding the two widths (16 feet) and the two lengths (22 feet) together.

16 plus 16 plus 22 plus 22 = 76 feet

STEP 2

Find the area of the walls to be painted by multiplying the perimeter from Step 1 (76 feet) by the height (10 feet).

76 feet times 10 feet = 760 square feet

STEP 3

Since you want two coats, you need to double the square feet from Step 2 (760 feet).

2 times 760 square feet = 1,520 square feet

STEP 4

To calculate how many gallons of paint you need, divide the total square feet from Step 3 (1,520) by the number of square feet covered by one gallon (400).

1,520 square feet divided by 400 = 3.8 gallons
Round up to 4 gallons.

You should buy more paint if the surface is rough or you are covering a dark color with a light color. Ask the clerk at the paint store to give you some guidance on how much more to buy.

STEP 5

To calculate your cost, multiply the cost of one gallon ($25) by the number of gallons you need from Step 4 (4).

4 gallons times $25 per gallon = $100

SOLUTION

You need 4 gallons, and it will cost $100.

In a Nutshell

HOW MUCH PAINT DO I NEED AND HOW MUCH WILL IT COST?

1. Calculate the perimeter of (distance around) the room.

    Perimeter = (2 × length) + (2 × width)

2. Find the area of the room.

    Area of room = perimeter × height

3. Calculate total area to be covered.

    Total area to be covered = area of room × number of coats

4. Calculate total gallons of paint needed.

    Gallons of paint = total area (in square feet) to be covered ÷ number of square feet covered by one gallon of paint (from the can label)

5. Calculate cost.

    Cost = gallons of paint × cost of one gallon

Spacing

In addition to calculating how much material you need, another math challenge for the homeowner is figuring out how to space things evenly. It is surprising how many projects require this skill—gardening, picture hanging, and putting up bookshelves, to name a few. Often, you might end up eyeballing where objects should go. If you are comfortable with that approach, just skip this section. But if you have too many holes in your walls from your guesswork, or you would just like some simple guidance, read on.

Spacing is actually very easy. You need to remember just one thing. When you place objects in a row, you create one more space than you have objects. Try it. Take a piece of paper and place two pens widthwise across the paper so that they are approximately equal distance from each other and the top and bottom of the paper. You will see that the two pens made three (2 plus 1) even spaces on the paper.

Plants in a Garden

Let's take a look at a common spacing problem: putting plants in a garden.

PROBLEM

It's a gorgeous spring day, and you have just spent a small fortune at Ned's Neighborhood Nursery. You are anxious to get your new pansies planted in your rectangular garden bed before the rain comes. You measure the bed and find that it is 12 feet long and 3 feet wide. You have 15 plants. How do you space the flowers evenly?

STEP 1

To center the flowers widthwise, divide the width of the garden (3 feet) by 2.

3 feet divided by 2 = 1.5 feet

■ STEP 2

1.5 feet isn't very meaningful, since your tape measure is in inches. Using the Home Improvement Changing Table on page 44, change feet into inches by multiplying the number of feet (1.5) by 12.

1.5 feet times 12 = 18 inches

1.5 feet does not mean 1 foot plus 5 inches. It means 1 foot plus .5 feet. To figure out how much .5 feet is, multiply .5 times 12 (inches in a foot) to get 6 inches. 1.5 feet equals 1 foot, 6 inches (or 18 inches).

■ STEP 3

To figure out how to center the flowers lengthwise, you first have to figure out how many spaces there will be in your garden. To do that, just add 1 to the number of plants you are planting (15).

number of spaces = 15 plants plus 1 = 16

■ STEP 4

Divide the total length of the garden bed (12 feet) by the number of spaces you have from Step 3 (16).

12 feet divided by 16 spaces = .75 feet

■ STEP 5

Again, you need to change feet into inches so you can work with your tape measure. Using the Home Improvement Changing Table on page 44, just multiply the number of feet from Step 4 (.75) by 12.

.75 feet times 12 = 9 inches

The lengthwise spacing for the flowers is every 9 inches.

■ SOLUTION

The flowers will be centered widthwise 18 inches from either side.

Lengthwise, the flowers will be spaced at even intervals 9 inches apart and 9 inches from each end.

Check the minimum spacing requirement on the little plastic tag stuck in the soil of your plants and flowers to be sure they will be far enough apart to have room to grow.

Shelving

To space shelves correctly, you use the same technique that you did with the flowers, but you only need to center them lengthwise. Just add 1 to the number of shelves you have, and divide this total into the length of your available space. That answer will tell you how far apart the centers of the shelves should be. For example, if you have five shelves and 12 feet of available space, divide 12 feet by six (5 plus 1) to arrive at two feet of spacing between shelves. Starting two feet from the top or the bottom, place the center of each shelf two feet from the last.

Hanging Pictures

The first thing you need to know about hanging pictures is that there is really no right or wrong way to do it. You can hang pictures wherever and however you choose. Here you will learn a very simple technique for hanging pictures at about eye level, centered at about 66 inches from the floor. If you want to go a little higher or lower, feel free. I tend to like my pictures hung a little lower, so I center around 62 inches.

PROBLEM

You have been taking art lessons and are quite pleased with your first piece, a sort of Cubist still life with neoclassic overtones. You decide you want to surprise your family by hanging it in a prominent place in the living room, and you want to center it just right. How do you do it?

STEP 1

Measure the width of the space you want to center the picture on (a wall, the space between two other pictures, the space between two pieces of furniture). In this case, it is an area next to the fireplace that is 60 inches wide.

60 inches

STEP 2

Figure out the center of the space by dividing by 2.

60 inches divided by 2 = 30 inches

STEP 3

Mark the center of the space with a small vertical line (that's up and down) 30 inches from one side.

STEP 4

Measure the height of the frame.

36 inches

STEP 5

Divide the height of the frame from Step 4 (36 inches) by 2 to find half its height.

Half of the frame's height = 36 inches divided by 2 = 18 inches

STEP 6

Hold the wire on the back of the picture up taut, and measure the distance from the top of the frame to the highest point of the wire.

3 inches

■ STEP 7

To calculate how high the picture hook should be, add half the frame's height from Step 5 (18 inches) to 66 inches, then subtract the wire measurement from Step 6 (3 inches).

66 inches plus 18 inches minus 3 inches = 81 inches

■ STEP 8

Draw a small horizontal line at the distance from Step 6 (81 inches) from the floor.

■ SOLUTION

Hang the bottom of the picture hook where the two (vertical and horizontal) lines intersect.

Hanging More Than One Picture

If you are hanging more than one picture, you can use the same strategy you used with one picture. For two pictures around the same size, just break your hanging area into two equal parts by dividing the width of your space by two. For example, if your hanging area is 10 feet wide, divide it into two equal spaces, each 5 feet across. Then just center the pictures in the newly created spaces, using the method above. For three pictures around the same size, divide the width of your space by 3, and follow the same procedure.

In a Nutshell

SPACING
General Spacing
1. Measure the length of the area in which you want to place your items.

continued

---

continued from previous page

2. Calculate the number of spaces that will be created by your items.

   Number of spaces = total number of items + 1

3. Calculate the length of the spacing between items.

   Length of spacing between items = length of the area (from #1) ÷ number of spaces (from #2)

4. Starting from either end of your space, place the center of each item at regular intervals equal to the length of the spacing from #3.

5. If the area you are covering has width as well (like a garden) center the items widthwise by placing them ½ the distance from either side.

Picture Hanging

1. Measure the width of the space where you will hang the picture.

2. Divide the space by 2 to find the center.

3. Mark a vertical line at the center.

4. Measure the height of the frame. Halve the height by dividing it by 2.

5. Measure the distance from the hanging wire, fully extended up, to the top of the frame.

6. Calculate the distance of the hook from the floor.

   Distance of the hook from the floor = 66 inches + ½ the frame height from #4 − the wire measurement from #5

7. Mark a horizontal line the distance of #6 from the floor.

8. Put the bottom of the picture hook where the two lines intersect.

While this chapter can't help you with uneven walls or smashed fingers, it should give you the confidence to tackle the many numbers involved in home improvement projects. And even if you end up hiring a contractor to do some of the work, you will be in a much better position to verify estimates for the amount of materials and the costs involved.

2 Math away from Home

4 Tipping

L et's start this chapter by debunking a great myth—that there is a right way to tip. Sure, our society has adopted a couple of tipping conventions, but who wants to be conventional?

If you're in a restaurant and the server was pleasant, prompt, and didn't spill anything on you that you can't get out, 15 percent is customary, but not required. A 20 percent tip is nice if the service was exceptional, but again, it's not cast in stone. On the other hand, if the meal was a buffet or the server was inattentive, it's okay to leave 10 percent. Also, remember that restaurants generally include a standard 15 percent tip or higher in the final bill for large parties or banquets.

There are, of course, lots of places other than restaurants where you tip—hair salons, barber shops, and cabs—all of which have their own confusing customs. In these places, your tip should reflect your level of satisfaction with the service, as well as how much cash you have in your wallet. If you get stuck, 10 to 20 percent works in all of these places.

Below I have included two approaches to tipping. The first is for you precise types who want to leave just the right tip and don't mind doing a little on-the-spot figuring or taking out the calculator in public. The second is a quick-and-dirty approach that will get you to pretty much the same place, and you can commit this approach to memory without much effort. If neither of these two approaches works for you, just pick up one of those little tip cards at the bookstore, stick it in your wallet, and forget about it.

Figuring Out the Exact Tip

Figuring out the exact tip is really pretty easy once you get the hang of it. It will require you to do a little arithmetic, however, either on your napkin or with that little calculator you stealthily keep in your purse or pocket. After you figure out what percentage tip you want to leave, there are just two steps. First change the tip percentage to a decimal by dividing by 100, then multiply it by your tab. For example, if you want to leave a 15 percent tip on a $26 tab, first change the 15 percent to .15 by dividing 15 by 100. Then multiply .15 by $26 to arrive at a tip of $3.90. Just round up to $4, and you are through. Let's look at an example.

PROBLEM

You are out to dinner with friends, and the check comes. Your generous but insincere "Let me get that . . ." is not met with the polite refusals you anticipate, and you are stuck with the bill. The tab, before tax, for you and your companions comes to $66.75. (With restaurant tax being what it is today, you should always tip on the pretax amount.) The service was good but not excellent, so you want to leave a standard 15 percent tip. How much do you leave?

STEP 1

Determine the percent tip you want to leave.

15%

STEP 2

To get the tip into decimal form, divide the percentage by 100.

15 divided by 100 = .15

An easy way to divide by 100 is to move the decimal place to the left two places. If there's no decimal place (as is the case above) just put one to the right of the number. If you can't visualize moving decimal places easily, just put your pencil on the original decimal place. Then skip the two numbers to your left, and put the new decimal place there.

STEP 3

Multiply the pretax tab ($66.75) by the tip percentage in decimal form from Step 2 (.15).

$66.75 times .15 = $10.01

STEP 4

Round the result of Step 3 ($10.01) to the nearest dollar.

$10.00

When rounding dollars, remember when the total ends in ".50" or more, go *up* to the next dollar. If the total ends in ".49" or less, go *down* to the dollar amount you have.

SOLUTION

Leave a $10.00 tip.

A Quick-and-Dirty Approach to Tipping

If you don't have a calculator, or you're too embarrassed to pull one out, there is a very easy way to quickly calculate a 15 percent tip. You can do this either in your head or on a napkin. (Only on a *paper* napkin, of course.)

PROBLEM

Let's take our previous example of leaving a 15% tip on $66.75.

■ STEP 1

Round the tab to the nearest $10.00.

$70.00

To round to the nearest value of $10.00: if the tab ends in 5.00 to 9.99, round *up* to the next value of $10.00; if the tab ends in less than 5.00, round off the $10.00 you have. In the example above, you round up to $70.00 because the ending of the tab, "6.75," is greater than 5.00. If the tab were $63.25, you would round down to $60.00, since the ending of the tab, "3.25," is less than 5.00.

■ STEP 2

To calculate 10%, take the result of Step 1 ($70.00) and move the decimal one place to the left.

$7.00

■ STEP 3

To calculate 5%, divide the result of Step 2 ($7.00) by 2.

$7.00 divided by 2 = $3.50

■ STEP 4

To get 15%, add the results of Step 2 ($7.00) and Step 3 ($3.50) (10% plus 5% = 15%).

$7.00 plus $3.50 = $10.50

■ STEP 5

Round the result of Step 4 ($10.50) to the nearest dollar.

$11.00

■ SOLUTION

Leave an $11.00 tip.

You can see by comparing the two methods that the quick-and-dirty approach is not as precise. If the tip seems low, just add a dollar or two. If it seems too high, subtract a dollar or two. Tipping does not have to be exact.

 For a 10% tip, just leave the result of Step 2. For a 20% tip, just double the result of Step 2.

In a Nutshell

TIPPING
Figuring Out the Exact Tip

Tip = (tip percent ÷ 100) × pretax tab

 Do the division first.

A Quick-and-Dirty Approach to Tipping

1. Round the pretax tab to the nearest $10.00.

2. Take the amount from #1 and move the decimal point one place to the left (e.g., $20.00 becomes $2.00).

3. 15% Tip = (result of #2) + ½ (result of #2).

For 10%, just use the result of #2.

For 20%, just double the result of #2.

One last thought on tipping. Some people can figure tips in their heads, some can't. If you can't, don't worry about it. Your brain is just made for more important stuff. And take heart. There are a lot more people out there faking their way through tipping than you think.

5 | Shopping

A h, shopping. Love it or hate it, you have to admit it is a great American pastime. Whether you adore it or abhor it, let's be honest—everyone's looking for a deal. I have a friend who actually bought her current home because of its proximity to an enormous outlet mall.

You have undoubtedly found that getting a bargain has gotten very complicated. Markdown percentages, promises of free merchandise, outlet malls, Internet buying, and all sorts of coupons confuse even the most seasoned consumer.

In this chapter we will walk through the most common shopping challenges you could encounter. We will go over markdown percentages—how to use them to figure out what the sale price is and how to calculate them. This chapter will also explain comparison shopping. In particular, it will describe how to use unit pricing and what to do when unit pricing is not available. Finally, the chapter will cover coupons and Internet shopping.

Markdown Percentage

Many stores use markdown percentages to indicate how big a savings is. Percent literally means *per centum,* or *per one hundred.* A percent is really just a number divided by 100. For example, 50 percent is .50 (50 divided by 100). Thirty-five percent is .35 (35 divided by 100).

In shopping, a percent is used to give you an idea of the proportion of the sale relative to the original price. Therefore, it is sometimes helpful to think of percents in terms of fractions. "Fifty percent off" means that the item is one-half off of its original price. "Twenty-five percent" off means it is one-fourth less. Use the chart below to give you an idea of some common percents and the fractions they represent.

Common Percents in Fraction Form	
Percent	Fraction
25	¼
33	⅓
50	½
75	¾

You Know the Markdown Percentage—What's the Sale Price?

Often when you're faced with a sale item in a store, the price tag has the original price listed, but the sign on the rack says to take a certain percent off. You are left wondering what the price of the item is. Let's run through an example of how to find out.

PROBLEM

You have been eyeing this great $120 sweater at Stacy's Upscale Boutique and have been waiting for Stacy's one and only annual sale to buy it. The sale is here, and everything is marked down by 30%. You want to know the sale price of the sweater.

STEP 1

Determine the percentage you actually pay by subtracting the markdown percentage (30%) from 100%.

100% minus 30% = 70%

STEP 2

To get the percentage you pay into a numeric form you can use, divide the percentage you actually pay from Step 1 (70) by 100.

70% divided by 100 = .70

To divide a number by 100 you can either use your calculator or just move the decimal place two places to the left. In the example above, you assume a decimal point after 70, so 70 is "70.". Moving the decimal point two places to the left produces ".70". If you can't eyeball moving decimal places, just place your pencil point on the decimal point, skip over the two digits to the left, and place the decimal point at that spot.

STEP 3

To find the sale price, multiply the percentage you pay (in decimal form) from Step 2 (.70) by the original price.

.70 times $120 = $84

SOLUTION

The sale price of the sweater is $84.

You Know the Sale Price—What's the Markdown Percentage?

Sometimes just a sale price is listed for an item. For example, you might see an ad for a washing machine that lists the original price at $500 and the sale price at $425. You might be curious, then, to know the markdown percentage in order to determine whether you are getting a good deal or not. The following example shows you how to calculate markdown percentage when you only know the original and sale price of an item.

PROBLEM

Your mattress is getting really lumpy, and your wife has already moved into the guest room. You don't want to buy a new mattress until you find a really good sale. You finally see an ad in the paper for a mattress marked down from $400 to $325 at Margie's Mattress Menagerie. It seems like a good buy, but you want to know the markdown percentage to see if the savings is really that good.

STEP 1

To find the actual markdown, subtract the sale price ($325) from the original price ($400).

$400 minus $325 = $75

STEP 2

Divide the actual markdown from Step 1 ($75) by the original price.

$75 divided by $400 = .19

STEP 3

To find the markdown percentage, multiply the result of Step 2 (.19) by 100.

.19 times 100 = 19%

 When you multiply a number by 100, you can use your calculator, or just move the decimal point to the right two places. You can either do that in your head or put your pencil point on the decimal point and move it to the right two digits.

■ SOLUTION

Markdown percentage is 19%.

(Go for it. It's a small price to pay for marital harmony.)

In a Nutshell

MARKDOWN PERCENTAGE

You Know the Markdown Percentage—What's the Sale Price?

1. Percent you pay = 100 − markdown percentage

2. Sale price = (percent you pay ÷ 100) × original price

 Do the division *before* the multiplication.

You Know the Sale Price—What's the Markdown Percentage?

1. Actual markdown = original price − sale price

2. Markdown percentage = (actual markdown ÷ original price) × 100

Comparison Shopping

In comparison shopping, you are simply trying to buy the product you want at the lowest possible price. It sounds easy enough, but the way products are packaged and marketed often makes it more complicated than it has to be. If you have a favorite brand or a strong preference for a type of packaging (sometimes it is just worth it to pay more for those convenient orange juice jugs), then you should just buy what you want and forget about comparing prices. If you really want to get the best price for a product, read on for a few tips.

Using Unit Pricing

Unit pricing, now available in many stores, is a fabulous way to compare the value of products. Simply put, unit pricing is the cost of buying one unit of a product. Many grocery stores now list for each product what it would cost to buy just one unit—say one ounce, one quart, or one sheet—so you can easily compare the cost of different brands or different sizes of a product. Unless you have a favorite brand or size, you just pick the product that has the smallest unit price. Remember that with items that come in great quantity, unit pricing is often based on a large number of items. For example, unit pricing on napkins and paper towels is usually done by hundred-count.

Unit pricing can use a variety of different measurements depending on the item and its price. Sometimes it's dollars per pound. Sometimes it's cents per ounce. In some cases, the unit pricing goes to tenths of a cent. Don't let the different measurements confuse you. On like items, unit pricing almost always uses the same measurement. Just choose the one with the lowest price and forget about it.

Comparing Prices When Unit Pricing Is Not Available

Unfortunately, the requirements for unit pricing vary from state to state, so it might not be available to you. Additionally, some stores do not provide unit pricing on sale items. And if you're using coupons, unit pricing doesn't really help you that much because it reflects the price before the coupon is deducted.

When unit pricing is not available, you have to calculate the per-unit price yourself. The following example shows you how to do that.

PROBLEM

Although it seems as if you just bought some, your kids tell you that you need ketchup. (You suspect food fights,

but there are just some things you'd rather not know.) While at the grocery store, you see that the 36-ounce store-brand ketchup costs $1.89, and the 40-ounce bottle of Big Red Ketchup is on sale for $2.79. You don't have any particular preference for either brand or size; you just want to get the best buy. Because the bottles contain different amounts of ketchup, you can't just buy the cheapest bottle, and unit pricing is nowhere to be found. Which bottle should you buy?

STEP 1

Start with the store-brand ketchup. Divide the cost of the store brand ($1.89) by the number of ounces in it (36) to determine what one ounce of the store brand would cost.

$1.89 divided by 36 ounces = $.053, or 5.3 cents, per ounce

To get from dollars to cents, either multiply dollars by 100, or just move the decimal point two places to the right. If you have trouble visualizing that, just put your pencil on the decimal point and skip over two numbers to the right, then place the decimal point right there. For example, $.53 equals 53 cents.

STEP 2

Now try Big Red. Divide the cost of Big Red ($2.79) by the number of ounces in it to determine what one ounce of Big Red would cost.

$2.79 divided by 40 ounces = $.07, or 7 cents, per ounce

(To get from $.07 to 7 cents, just multiply $.07 by 100.)

STEP 3

Compare the cost of one ounce of each of your choices.

Store brand 36-ounce = 5.3 cents per ounce

Big Red 40-ounce = 7 cents per ounce

SOLUTION

The 36-ounce store brand is the better value at 5.3 cents per ounce.

Comparing Prices for Foods That Come in Different Containers

Even if unit pricing is available, it's not very helpful if you are comparing the prices of foods that come in different kinds of containers. For example, the unit pricing for a can of iced tea mix will be based on an ounce of powdered mix, not on an ounce of iced tea actually made. On the other hand, the unit pricing for bottled iced tea will be based on one ounce of tea ready to drink. So, if you want to compare the prices of the iced tea mix and the bottled iced tea, unit pricing will not help you.

To compare prices for foods that come in different types of containers, you have to compare the unit cost of both products made into their final form.

PROBLEM

You love the convenience of buying orange juice in those big jugs, but you suspect you are paying a large price for this convenience. The gallon jug of orange juice sells for $6.29; the unit pricing on the store shelf says it costs 4.9 cents per ounce. The frozen orange juice sells for $1.59, and the package says it makes 48 ounces of orange juice. Which is the better buy?

Since you know the unit pricing for the gallon jug of orange juice is 4.9 cents an ounce, you need to figure out how much the frozen orange juice costs for 1 ounce *prepared* juice. The unit pricing for the frozen orange juice isn't helpful because it is based on ounces of concentrate, *before* the orange juice is made.

▌STEP 1

Let's start with the frozen concentrate. Calculate how much one ounce of orange juice made from the frozen concentrate costs by dividing the price ($1.59) by the number of ounces it makes (48).

$1.59 divided by 48 ounces = $.033, or 3.3 cents, an ounce

(You just multiply $.033 times 100 to get 3.3 cents.)

▌STEP 2

Compare the jug orange juice at 4.9 cents an ounce with the prepared frozen orange juice at 3.3 cents an ounce.

The frozen orange juice is less expensive at 3.3 cents an ounce for prepared juice.

▌SOLUTION

Buy the frozen orange juice.

Buy One, Get One Free

"Buy one, get one free," or some variation of this slogan, is a very popular way of marketing products. Marketers know that it is enticing to the consumer to get a free item. Despite the fact that you may indeed be getting something free, this is not always the cheapest way to buy a product. Just as in the examples above, you should figure out what one item actually costs and then do the comparison. To see this point illustrated, look at the example below.

▌PROBLEM

An ad in this morning's paper catches your eye. Doug's Discount Drugs is having a sale on Pearlywhite toothpaste, which happens to be your favorite. Buy one 8-ounce tube

at $3.75 and get one free. That sounds pretty good. You see another ad for Wilma's Wacky Warehouse, where the same Pearlywhite 8-ounce tube is on sale for $1.75 each. Which is the better buy?

What you really want to do is figure out what one tube of toothpaste costs both at Doug's and at Wilma's. You can then compare tube to tube, since the tubes are the same 8-ounce size.

STEP 1

Calculate the real cost of one tube of toothpaste at Doug's by dividing the original cost of one tube ($3.75) by 2 (since you get 2 tubes).

$3.75 divided by 2 = $1.88 per tube

STEP 2

Compare the real cost of one tube at Doug's ($1.88) with the cost of one tube at Wilma's ($1.75).

A tube of toothpaste at Wilma's is cheaper at $1.75.

SOLUTION

At $1.75 for one tube, Wilma's is offering the best price.

You can see by this example "buy one, get one free" does not always offer a bargain. Even though you don't technically "get one free" at Wilma's, it's still cheaper to buy your toothpaste there.

Sometimes you can do these comparisons easily in your head by making some commonsense judgments. Looking at the example above, you might think to yourself, I could buy two tubes of toothpaste at Wilma's for $3.50 (2 times $1.75). The same two tubes would cost $3.75 at Doug's, so Wilma's is clearly the better buy.

You can use the same logic in a "buy *two*, get one free" scenario. Just calculate the real cost of one item. For example, consider a "buy two, get one free" sale on mops, with each original mop costing $5.99. You round the mop price to $6.00 and calculate that you would have to spend about $12.00 to get the two mops required. However, since you get one free, you really get *three* mops (2 plus 1) for $12.00, so each one really costs $4.00 ($12.00 divided by 3). Of course, in this "buy two, get one free" sale, you need to decide if you really need all those mops.

Coupons

I have a confession to make. I don't use coupons. I can't keep them straight or organized. When I go to use them, they are always expired. But remember my friend who bought her house near the outlet mall? She swears she saves $100 a month on family groceries by using coupons.

There are a number of different kinds of coupons these days. If the coupon is for a specific amount of money off, like $10.00 off an oil change at Slick Dick's service station, then just hand in the coupon and make sure Dick gives you the discount. If the coupon takes a certain percent off an item or off an entire purchase, use the markdown percentage example on page 87. Just treat the markdown percentage on the coupon as if it were the store markdown.

Grocery store coupons have gotten particularly complicated, with some stores offering double or triple value on certain days of the week, half off on some items, special values for store cardholders, and even mail-in rebates. So, to compare prices of products at the store you have to first deduct all discounts from the price of the product. Then use the comparison shopping examples on page 88 to figure out what your real unit cost is.

PROBLEM

You are making your weekly assault on the grocery store. You are armed with a coupon for 75 cents off Cuddly Clothes fabric-softener sheets. The original price of the Cuddly Clothes brand is $4.99 for the 120-count (120 sheets in the package). But your "frequent shopper" card gives you the store brand's 200-count fabric-softener sheets for $7.89, which is unit-priced at 3.9 cents per sheet. Which product offers the better deal?

STEP 1

Start with the Cuddly Clothes brand. To determine the discounted price, subtract the 75-cent coupon discount from the original price ($4.99).

$4.99 minus $.75 = $4.24

(Here, you need to convert 75 cents to a dollar equivalent, $.75, so you can use your calculator more easily.)

To convert cents to a dollar amount, either divide by 100 or just put a decimal point at the end of the cents and move the point over two places to the left. Be careful, though. If you're converting something like 4 cents to a decimal, there aren't two places to move, so you have to put a "0" in front of the 4 before you do anything else (4 cents = 04. cents = $.04).

STEP 2

To figure out how much one sheet of Cuddly Clothes costs, divide its discounted price from Step 1 ($4.24) by the number of sheets in the box.

$4.24 divided by 120 = $.035, or 3.5 cents, for one sheet

(Here, just multiply by 100 to go from dollars to cents, or move the decimal point over two places to the right.)

STEP 3

Compare the 3.9-cent unit price for the store-brand fabric softener with the 3.5-cent price per sheet of the discounted Cuddly Clothes.

The Cuddly Clothes with your coupon is the cheaper product.

SOLUTION

Buy the Cuddly Clothes brand.

Internet Shopping

It's hard to beat the convenience of Internet shopping. Any of the shopping techniques discussed above can be applied to shopping on-line. Just make sure you include any costs for processing and shipping your order in your price comparisons.

In a Nutshell

COMPARISON SHOPPING

Comparing Prices When Unit Pricing Is Not Available

1. Calculate the cost of one unit of each product you are comparing.

    Cost of one unit of a product = total price of a product ÷ the number of units in the package

2. Buy the product with the lowest price.

Comparing Prices for Food That Comes in Different Containers

1. From the package directions, figure out how many units of the product each container makes (e.g., iced tea mix might make 64 ounces).

continued

continued from previous page

2. Calculate the unit price for the prepared product from each type of container.

Unit price for a container = price of product ÷ the number of units that container makes

3. Compare unit price for each container, and buy the cheapest.

Buy One, Get One Free

1. Calculate the true cost of buying one product.

True cost of buying one product = original cost of one product ÷ 2 (since you get two products for the price of one)

2. Compare the true cost of one product in the "buy one, get one free" scenario to the cost of alternative products. Buy the cheapest one.

Buy Two, Get One Free

1. Calculate the total cost of this offer by multiplying the cost of one item by two since you have to buy two items.

Total cost of offer = cost of one item × 2

2. Calculate the true cost of buying one product.

True cost of buying one product = total cost of offer from #1 ÷ 3 (since you get three products)

3. Compare the true cost of buying one product in the "buy two, get one free" scenario to the cost of alternative products. Buy the cheapest one.

Coupons

Deduct the cost of the coupon from the original price of the product, then compare the price of the product using one of the methods above.

At the end of the day, shopping can get pretty complicated. For some, making your way through all the discounts and coupons and sales is an entertaining game. For others, it's sheer drudgery. In either case, you are now armed with some skills that should make shopping a lot easier. And as my friend who lives near the outlet mall always says, "Shop on."

6 | Traveling

W hat could be more glorious than traveling? Places to explore. Adventures to have. Old friends to reconnect with. No time schedules, no billable hours, no constraints. Just the open road ahead of you, and . . . Dad, when are we going to be there?

This chapter will answer that question as well as several others you might encounter while traveling. The chapter starts with some hints on how to deal with time differences while traveling. Then it moves on to simplify such traveling issues as computing gas mileage, calculating average speed, and determining how far can you go in a given period of time. You will learn how to convert kilometers to miles and miles to kilometers and learn how to change temperature from Celsius to Fahrenheit. You will also acquire some easy techniques for converting foreign currency.

Working with Time

Many traveling questions involve understanding how time works. For some people, it is very natural to work with sixty-minute hours and sixty-second minutes, and they go back and forth from A.M. to P.M. without a problem. But for many, working with time is complicated and confusing. If you are among these people for whom working with time is difficult, take heart. The next few pages will explain some easy ways to deal with tricky time problems.

Time Intervals

Often when traveling you need to work with intervals of time. You might need to calculate how long a trip took, or when you expect to be at your destination. The thing to keep in mind is that you can subtract and add times just like you would any other numbers. You just need to make a couple of adjustments. Let's take an example.

PROBLEM

You are supposed to pick up your daughter at Bambi's Beauty College at 11:30 A.M. Allowing for some traffic, you think it will take you about 2 hours and 45 minutes to drive there. You are on a tight schedule. When do you have to leave to get to Bambi's on time?

STEP 1

To calculate when you have to leave, subtract the traveling time from the destination time.

11:30

–2:45

It would be nice if you could subtract hours from hours and minutes from minutes, but you can't do that since the minutes in the bottom number (45) are larger than the minutes in the top number (30).

STEP 2

Just like in conventional subtraction, you need to "borrow" some minutes from the hours column. First, you take 1 hour from the 11 in the top number, leaving 10 hours. Then replace the one hour you took by adding the equivalent of 1 hour—60 minutes—to the minutes column of the top number (30). The problem becomes:

10:90 (60 plus 30 = 90)

−2:45

Now you can subtract both the hours and the minutes.

8:45

SOLUTION

You need to leave at 8:45 A.M.

Let's look at the above example a different way to demonstrate how to *add* time. Let's say you wanted to leave your house at 8:45 A.M. You know that it takes two hours and 45 minutes to get to Bambi's. What time will you get there? Set the problem up as you would any other addition problem. Add the hours to the hours and the minutes to the minutes.

8:45

+2:45

10:90

Since 10:90 is not a real time, you need to change it into something that makes sense by borrowing in the same way as you did in the subtraction problem above. This time you will take sixty minutes from the minutes column and give it to the hours column in the form of one hour. The minutes column becomes 30 (90–60) and the hours column becomes 11 (10 plus 1).

10:90 = 11:30

You will arrive at Bambi's at 11:30 A.M.

A.M. and P.M.

If you are both starting and finishing your trip in either the A.M. or P.M., it is relatively easy to work with time intervals. But some people have real trouble with time intervals if they travel between A.M. and P.M. If you are one of these, don't worry about it. Just use a twenty-four-hour clock to help you out.

Let's say you need to be at your sister's house at 4:00 P.M. It's 11:00 A.M. now, and you want to know how much time you have before you have to be there. To use the twenty-four-hour clock, just add twelve hours to the P.M. time (4:00 P.M. plus 12 hours = 16:00). Leave the A.M. time alone. Then just subtract the current time from the time you are due at your sister's (16:00 minus 11:00 = 5 hours) to find that you have five hours before you have to be there.

If you are traveling overnight, and therefore traveling from P.M. in one day to A.M. in another, you use a similar approach. Look at the following example.

PROBLEM

You are traveling with your small children to Grandma's house and prefer, for obvious reasons, to travel when they are sleeping. You plan to leave at 8:00 P.M. and you know that the drive is about 10 hours with stops. When will you arrive at Grandma's house?

STEP 1

Convert your starting P.M. time (8:00) to the 24-hour clock by adding 12 hours to it.

8:00 P.M. plus 12 = 20:00

STEP 2

Add your travel time (10 hours) to your start time from Step 1 (20:00) to calculate your arrival time in terms of the 24-hour clock.

20:00 plus 10 hours = 30:00

STEP 3

To allow for the fact that you are arriving on the next day, subtract 24 hours.

30:00 minus 24 hours = 6:00

SOLUTION

You will arrive at 6:00 A.M.

Calculating What Time It Is in Different Time Zones

While a pretty simple concept, time zones can get complicated quickly. Even if you know that there is a three-hour time difference between you in New York and your son in Los Angeles, you might not know how to apply that difference. Do you add it to or subtract it from your time? Complicating matters is the fact that some places honor daylight saving time and others do not.

You Know the Time Difference

If you know the time difference between two locations, the quick trick to figuring out whether you add or subtract the difference is to think about where your time zone is relative to the other person's zone. If you are east of that person, you *subtract* the difference *from your time* to calculate his time. If you are west of the other time zone, you *add* the difference *to* your time. Always orient to yourself. Check out the handy table below that summarizes this information for you:

Time-Zone Changer	
If you are:	**then:**
East of the other person	subtract the time difference.
West of the other person	add the time difference.

Let's see how this works.

PROBLEM

You are setting up a conference call in your office in Chicago for 10:00 A.M. Wednesday morning. You are including people who work in Washington, D.C., which you know

is 1 hour different than you, and from Seattle, which is 2 hours different than you. What time should you put on the meeting notice for the Washington, D.C., and Seattle participants?

STEP 1

Let's start with the D.C. people. Use the Time-Zone Changer above to see that, since you are west of Washington, D.C., you *add* the time difference (1 hour).

10:00 A.M. plus 1 hour = 11:00 A.M.

STEP 2

Let's move on to Seattle. Use the Time-Zone Changer above to see that since you are east of Seattle, you *subtract* the time difference (2 hours).

10:00 A.M. minus 2 hours = 8:00 A.M.

SOLUTION

Schedule the meeting for 8:00 A.M. in Seattle, 10:00 A.M. in Chicago, and 11:00 A.M. in Washington, D.C.

You Don't Know the Time Difference

What do you do, however, if you do not know the time difference, or if you are a little directionally challenged and could not tell east from west if your life depended on it? Not to worry. By simply searching the Internet for the words "time zone," you will have your choice of several free, easy-to-use sites that do all the math for you. On most of these sites you can just type in a city name, and the site will not only tell you what time it is there now but will give you the local time for any time and date you enter. In the above example, the person setting up the conference call could have typed in "10:00 A.M. in Chicago" and a certain date and checked what time it would be in Washington, D.C., and Seattle.

Many of these sites, however, do the calculations for you but do not give you the time difference per se. There are times, particularly when you are traveling, when you will want to know the actual time difference. For example, you might anticipate the need to make calls while you are on a trip and would like to make them at a civilized hour. To use an Internet time-zone chart to calculate time difference, you need to check what the time will be on a particular date in both your home and destination cities. Then, you will need to calculate the difference between them. Let's see how this works.

PROBLEM

You are planning your dream trip to Paris for April 2. You live in Salt Lake City, Utah. You have given the dog sitter your itinerary and want to be able to inform him when he might call you in case Rover needs something. How do you find the time difference?

STEP 1

Search the Internet for a time-zone site by typing "time zones" on your Web browser.

STEP 2

When you find an appropriate site, tell it you want to know the time in your home city (Salt Lake City) when it is 12:00 noon at your destination city (Paris). (Starting with 12:00 noon just makes your calculations easier. You could compare any time.)

Paris, April 2	**12:00 noon**
Salt Lake City, April 2	**3:00** A.M.

When calculating time differences, be sure to check what the time will be on the dates you plan to be away, since most of the world does not go on daylight saving time.

STEP 3

Calculate the time difference by subtracting the earlier time (3:00 A.M.) from the later time (12:00 noon).

12:00 noon minus 3:00 A.M. = 9:00 hours

STEP 4

When you are in Paris, you will be *east* of your dog sitter. Use the Time-Zone Changer table above, then, to see that when you are in Paris you will *subtract* 9 hours to get the time in Salt Lake City. When you are in Paris, the dog sitter will be *west* of you. Use the Time-Zone Changer table above to see that your dog sitter would *add* 9 hours to his time to get the time in Paris.

SOLUTION

The time difference between Salt Lake City and Paris is 9 hours.

When you are in Paris, subtract 9 hours to get the time in Salt Lake City.

Your dog sitter in Salt Lake City should add 9 hours to get the time in Paris.

Even if you do not have access to the Internet, you'll find most travel agents and relevant travel books have information on the time zone of your destination and how it relates to yours.

Another way to calculate domestic time differences is to use the time-zone map included in most major phone books. The map usually breaks the United States into six time zones with little clocks on top of each. To find the time difference between you and someone else, find your time zone and its clock, and calculate how many hours you add or subtract to get to the time zone of interest. If you live in Dallas, Texas, for example, you might want to compute the time difference between you and your colleague who works in Boise, Idaho. Find the clock on the time-zone map for your area. It might say 3:00 P.M. Then find the time-zone clock for your colleague's zone in Boise, which would say 2:00 P.M. Subtract the earlier time from the later time to get the time-zone difference (3:00 minus 2:00 = 1 hour). Use the Time-Zone Changer table on page 103 to see that since Dallas is *east* of Boise, you *subtract* the time difference (1 hour) from your time to get the time in Boise.

Traveling through Time
Another math time challenge is figuring out time intervals when you are traveling through different time zones. This skill is particularly important when dealing with airline schedules. Because of time-zone changes, you could actually arrive at your destination in one time zone at a time earlier than the time you left another time zone. To learn how to compute time intervals through different time zones, look at the following example.

PROBLEM

You are traveling from New York City to Denver, Colorado. Your flight leaves New York at 10:30 A.M. eastern time and arrives in Denver at 12:40 P.M. You know there is a 2-hour time difference between New York and Denver. How long is your flight?

STEP 1

First calculate the time it will be in New York when you arrive in Denver (12:40 P.M. Denver time). To do this, use the Time-

Zone Changer table on page 103 to see that, since New York is *east* of Denver, you *add* the time difference (2 hours).

12:40 P.M. plus 2 hours = 2:40 P.M.

■ STEP 2

Subtract your time of departure (10:30 A.M.) from your time of arrival (2:40 P.M.), all in New York time.

2:40 P.M. minus 10:35 A.M. = 4 hours and 5 minutes

As described above, you could use the 24-hour clock to help you with this problem if you have trouble going from A.M. to P.M. You do this by adding 12 hours to any time that is in the P.M. In the example above, 2:40 P.M. would become 14:40 (2:40 plus 12:00). The subtraction would then become 14:40 minus 10:35 (= 4 hours and 5 minutes).

■ SOLUTION

Your flight will take 4 hours and 5 minutes.

In a Nutshell

WORKING WITH TIME
Time Intervals

1. Add or subtract hours and minutes.

2. In subtraction, if the minutes in the earlier time are more than the minutes in the later time, "borrow" an hour from the later time, and add 60 minutes to the later time's minutes. Then subtract hours from hours and minutes from minutes.

3. In addition, if the minutes in the total exceed 59, change extra minutes into hours by subtracting 60 from the minutes column and adding an extra hour to the hours column.

continued

continued from previous page

4. To add and subtract time between A.M. and P.M., convert the P.M. time to the 24-hour clock by adding 12 to it. Then add or subtract as explained above.

Calculating What Time It Is in Different Time Zones

1. Use the Time-Zone Changer table on page 103 to determine whether you add or subtract the time difference, which depends on where you live.

2. If you don't know the time difference, consult the time-zone map in your phone book, or check an Internet site.

3. To calculate your travel time if you'll be crossing different time zones, first find out what your arrival time will be in the time zone of your *departure* city, using the Time-Zone Changer table on page 103. Then subtract your departure time from your arrival time to calculate travel time.

Driving

Now that you know how to travel through time, let's take a look at how most of us actually do travel—in our cars.

Fuel Efficiency—Computing Miles per Gallon

Because gas prices are high and because of environmental concerns, many people are interested in fuel efficiency these days. Fuel efficiency is generally measured in miles per gallon (mpg)—the number of miles you can go, on average, on one gallon of gas. The higher the mpg, the farther you can go on a gallon of gas. Accordingly, most new-car sales lists specify the car's average mpg, and many new cars actually compute the mpg as you go along. If you are interested in calculating the mpg for your car, take a look at the following example.

PROBLEM

When you got your first job, your parents gave you their big family car. It runs well, and you can fit half your office in the back seat, but it seems to use a lot of gas. You're concerned about gas prices and also about the car's impact on the environment. How do you figure out how many miles to the gallon you're getting?

STEP 1

Fill the car up with gas, and note the mileage on the odometer. Or, if the car has a trip meter, set it to "0."

odometer reading = 67,532

trip meter = 0

STEP 2

Next time you fill up with gas, note the new mileage on the odometer or trip meter and the number of gallons it takes to fill the tank.

odometer reading = 67,811

trip meter = 279

gallons to fill up = 19.7

STEP 3

If you do not have a trip meter, subtract the first odometer reading (67,532) from the second (67,811) to determine how many miles you drove on that tank of gas.

67,811 miles minus 67,532 miles = 279 miles

STEP 4

To calculate miles per gallon, divide the miles traveled on that tank from the trip meter or from Step 3 (279) by the number of gallons to fill up from Step 2 (19.7).

279 miles divided by 19.7 miles = 14 miles per gallon

SOLUTION

Your average miles per gallon is approximately 14.

Miles per gallon varies a lot between city and highway driving. Generally, it is far lower if you're driving stop-and-go than if you're driving on the highway at a consistently high speed.

How Long Will It Take to Get There?

Often when you plan a trip you need to figure out about how long it will take to get to your destination. That calculation allows you to plan overnight stays, meal stops, etc. To calculate total traveling time, just divide the total miles by the average speed you think you will go. Add extra time for stops and unexpected delays. (When I was driving in Idaho a few years back, I was held up for two hours by some stubborn sheep that decided they wanted to hang out in the middle of a major highway.)

The calculation of traveling time might involve fractions and decimals. This can get a little confusing. For example, 8.20 hours does not mean eight hours and twenty minutes. It means eight hours and .20 of an hour, which is actually twelve minutes (.20 hours times 60 minutes in an hour = 12 minutes.)

The Travel Time Changing Table below should help you with these calculations by giving you various times in three different forms: minutes, hours in fractions, and hours in decimals.

Changing Table

TRAVEL TIME

Minutes	Hours in Fractions	Hours in Decimals
15	¼	.25
30	½	.5
60	1	1.0
90	1½	1.5
120	2	2.0
180	3	3.0

The example below will take you through how to calculate traveling time and how to use this table.

PROBLEM

You are taking the kids on their first trip to Wally World, and everyone's excited. It's about 450 miles from your house, and you are going to try to make it in one day. You think you will average about 55 miles per hour while you're on the road and will have to take two 45-minute rest stops. How long will it take you?

STEP 1

Begin by computing driving time in hours by dividing the distance you expect to travel (450 miles) by the rate of speed (55 miles per hour) you expect to travel.

450 miles divided by 55 miles per hour = 8.2 hours

Round to 8 hours.

In rounding, you need to look at the number to the right of the place to which you are rounding. For example, when rounding 8.2 hours to whole hours, you look at the 2, since it is directly to the right of the hours place. If the number to the right is 5 or above, round up to the next hour. If it is less than 5, round down to the same hour. Since 2 is less than 5, you round 8.2 down to 8 hours.

STEP 2

Compute rest time by adding the time for the two rest stops (45 minutes each) together.

45 minutes plus 45 minutes = 90 minutes

STEP 4

Change minutes of rest time (90) to hours so you can more easily add driving time and rest time together. Use the Travel Time Changing Table on page 111.

90 minutes = 1.5 hours

STEP 3

Calculate total travel time by adding driving time from Step 1 (8 hours) and rest time from Step 2 (1½ hours).

8 hours plus 1.5 hours = 9.5, or 9½, hours

SOLUTION

You will be traveling approximately 9½ hours.

(Whew. That's a long day.)

How Far Can You Go in One Day?

Most likely, you know your family's tolerance for time spent driving in one day. Let's take another look at the example above. This time, let's set a specific limit of how many hours you want to travel in one day.

PROBLEM

You are taking the kids to Wally World, which is 450 miles from your home. You think you will average 55 miles an hour and only want to travel 6 hours in one day, with one 45-minute rest stop. How far can you go in 1 day? How many days will it take you?

STEP 1

Calculate actual driving time by subtracting rest time (45 minutes) from total travel time desired (6 hours).

6 hours minus 45 minutes = 6 hours minus .75 hour = 5.25 hours

(Here, you use the Time Travel Changing Table on page 111 above to substitute .75 hour for 45 minutes.)

STEP 2

Calculate driving distance in one day by multiplying driving speed (55 mph) by the number of hours you will drive in one day from Step 1 (5.25).

55 mph times 5.25 hours = 289 miles (approximately)

STEP 3

Calculate the number of days for the trip by dividing the total trip distance (450 miles) by the number of miles you will drive in one day from Step 2 (289 miles).

450 miles divided by 289 miles per day = 1.6 days, or a little more than a day and a half (since .5 = ½, and .6 is a little more than .5)

SOLUTION

You can drive about 289 miles in one day, and it will take you a little more than a day and a half to get there.

Average Speed

Many times you will want a sense of the average speed you travel at so you can plan future trips better. Let's look at our Wally World example one more time. But let's say you just set out and kept driving until you got there. You want to know what your average speed was.

PROBLEM

You have arrived at Wally World. Your odometer shows that it is about 450 miles from your house. You stopped for two 45-minute rest stops, and it took you about 9½ hours (or 9.5 hours) to get there. What was your average speed?

STEP 1

Calculate the time spent in the two rest stops (45 minutes each) in hours.

45 minutes plus 45 minutes = 90 minutes = 1.5 hours

(Use the Travel Time Changing Table on page 111 to see that 90 minutes equals 1.5 hours.)

STEP 2

To calculate actual driving time, subtract the time spent in rest stops from Step 1 (1.5 hours) from the total travel time (9.5 hours).

9.5 hours minus 1.5 hours = 8 hours

STEP 3

To calculate average speed, divide the total distance (450 miles) by the number of driving hours from Step 2 (8 hours).

450 miles divided by 8 hours = 56 miles per hour

SOLUTION

You averaged about 56 miles per hour.

(The figures vary a little between this example and the first one because you did some rounding in the first example.)

In a Nutshell

DRIVING

Fuel efficiency (miles per gallon) = miles traveled ÷ number of gallons used

How long will it take to get there? (time spent driving) = distance traveled (in miles) ÷ average speed (in miles per hour)

How far can you go in one day (in miles)? = average speed (in miles per hour) × time spent driving in one day (in hours)

Average speed (in miles per hour) = distance traveled (in miles) ÷ time spent driving (in hours)

If you need to change hours to minutes, or minutes to hours, just use the Travel Time Changing Table on page 111.

Foreign Travel

While foreign travel can be exciting and fun, there are some additional hassles to traveling in another country. Unfortunately, this chapter won't help you much with language barriers, jet lag, and lost luggage, but it can help you solve most common math challenges. In particular, you will learn how to conquer some driving challenges, like changing kilometers to miles and liters to gallons. If temperatures in Celsius give you a chill, keep reading for an explanation of how to convert Celsius to Fahrenheit. And finally, you will have a step-by-step explanation of the thorniest of travel problems, converting currency.

Driving Abroad

Changing between Miles and Kilometers

Many countries you may travel in use the metric system to measure distance. Mileage signs on the roadway will be in kilometers, and your odometer will most likely read kilometers per hour instead of miles per hour. It can be a little daunting to be cruising along some beautiful country road in Belgium and look down to see that you are going 100! To get a handle on kilometers, you will need to convert kilometers to miles, which is easy to do with a simple formula. You can use another simple formula to go from miles to kilometers.

Since it won't always be convenient to whip out your calculator, the examples below walk you through easy ways to change kilometers to miles and miles to kilometers, in your head. For those times when you need a more precise number, there are also examples of how to convert exactly between miles and kilometers. Let's first look at the more common problem: changing kilometers to miles.

Kilometers to Miles—Quick-and-Dirty

PROBLEM

You have successfully negotiated airports, customs, lodging, and renting a car. You are now happily cruising down the Autobahn in Germany looking for signs for your destination—Gukkendorf. You see a sign ahead that says Gukkendorf 128 kilometers. How far is that in miles?

STEP 1

Since you don't want to pull out your calculator while you're racing down the Autobahn, you do an approximation by first rounding up the number of kilometers (128).

130 kilometers

(You rounded up to 130 so you could have a number you could work with in your head. Some people can't hold onto numbers in their heads. Don't despair. Keep rounding until you get a number you can work with. Here, you could have used 150 or 100 for 130 if you needed to. Just remember if you round up, your estimate will be a little high. If you round down, it will be a little low.)

STEP 2

Multiply the number of kilometers from Step 1 (130) by 6.

130 times 6 = 780

STEP 3

Place a decimal after the result of Step 2 (780) and move it one place to the left, to get an estimate for how many miles are in 130 kilometers.

780. moved one decimal point to the left = 78.0, or 78, miles

SOLUTION

It is approximately 78 miles to Gukkendorf.

Kilometers to Miles—Precise

PROBLEM

You have approximated the distance to Gukkendorf to be about 78 miles. Your husband, who is not driving, offers to use his calculator to do a more precise calculation for you. What should he do?

STEP 1

Multiply the exact number of kilometers (128) by .625.

128 times .625 = 80 miles

You can also multiply by ⅝ instead of .625 if that's any easier for you. Believe it or not, some people actually like fractions better than decimals.

SOLUTION

It is 80 miles to Gukkendorf.

(Looking at the example above, you can see that your approximation of 78 miles was pretty close.)

Miles to Kilometers—Quick-and-Dirty

PROBLEM

You are on a wonderful biking vacation through the vineyards of France. You are riding along when a French cyclist coming your way stops and asks you how many kilometers it is to the Vinichy Vineyard. Thrilled that your high school French has served you well enough to actually understand her, you would like to help. You know you passed the vineyard about 8 miles back, but how many kilometers is that?

STEP 1

Since you don't have a calculator with you, you decide to use a quick-and-dirty approach. Start by dividing the number of miles (8) by 2.

8 divided by 2 = 4

STEP 2

Add the original number of miles and the result of Step 1 (4) together. (What you're actually doing here is multiplying the number of miles by 1½.)

8 plus 4 = 12

SOLUTION

The Vinichy Vineyard is about 12 kilometers down the road.

Miles to Kilometers—Precise

PROBLEM

Your biking partner remembers that she has a calculator function on her high-tech odometer, and you decide you want to give the French bicyclist a more precise estimate of distance. If you are correct that it is about 8 miles back to the Vinichy Vineyard, how many kilometers is that?

STEP 1

Multiply the number of miles (8) by 1.6.

8 miles times 1.6 = 12.8 kilometers

SOLUTION

It is 12.8 kilometers to the Vinichy Vineyard.

Changing between Gallons and Liters

Many countries use liters instead of gallons as a measurement for gasoline. To deal with liters, it is helpful to know the equivalent number of gallons. For example, if the store clerk asks you how much milk you want to purchase, and the milk is sold in liters, it is good to know that one liter equals about ¼ gallon, or approximately one quart. As is the case with miles and kilometers, most of the time an approximate conversion is sufficient. In these instances, use the quick-and-dirty techniques below. An example is also included of a more precise conversion.

Liters to Gallons—Quick-and-Dirty

PROBLEM

You are driving along a beautiful scenic highway in the Canadian Rockies. You are nearing empty, so you stop at

a gas station to fill up. You fill up your tank, and it takes about 42 liters. How many gallons did it take?

STEP 1

Round the liters (42) to 40 so you can more easily deal with the number in your head.

STEP 2

Divide the number of liters from Step 1 (40) by 4, since there are about 4 liters in a gallon.

40 divided by 4 = 10

SOLUTION

Your tank took about 10 gallons.

Liters to Gallons—Precise

PROBLEM

Your daughter is getting a little bored by all the travel, so you suggest she use her calculator to give you a more precise calculation of how many gallons it took to fill up your tank at the Canadian gas station. Your tank took 42 liters. How many gallons is that?

STEP 1

Multiply the number of liters (42) by .264.

42 liters times .264 = 11.1 gallons

SOLUTION

Your tank took 11.1 gallons of gas.

It is unlikely that you will have to change from gallons to liters while traveling. If you need to convert gallons to liters, simply multiply the number of gallons by 4.

DRIVING

Changing between Miles and Kilometers

Kilometers to Miles—Quick-and-Dirty

1. Multiply the number of kilometers by 6.

2. Take the result of #1, and move the decimal one place to the left.

Kilometers to Miles—Precise

Number of miles = number of kilometers × .625 *or* number of kilometers × ⅝

Miles to Kilometers—Quick-and-Dirty

1. Divide the amount of miles by 2.

2. Kilometers = miles plus result of #1.

Miles to Kilometers—Precise

Number of kilometers = number of miles x 1.6

Changing between Gallons and Liters

Liters to gallons—Quick-and-Dirty

Number of gallons = number of liters ÷ 4

Liters to Gallons—Precise

Number of gallons = number of liters × .264

Temperature—Fahrenheit and Celsius

Now that you have figured out how to measure distance while traveling abroad and how to fill up your gas tank, let's move on to the weather—always a safe subject. Often when traveling you will find that the temperature is measured on the Celsius scale rather than Fahrenheit. To get an idea of what the actual temperature is, you will want to be able to change the Celsius temperature to Fahrenheit. Changing from one to the other is really pretty easy. There are a couple of ways to do it, depending on how precise you want to be. For a general guide, just use the handy changing table below.

Changing Table

CELSIUS/FAHRENHEIT

	Fahrenheit	Celsius
awfully cold	0	-18
freezing	32	0
cool	50	10
beautiful	70	21
a real scorcher	105	41
boiling water	212	100

If you need a little more precision than the Celsius/Fahrenheit Changing Table offers you, you will find examples below of how to change between Celsius and Fahrenheit. As in other traveling conversions discussed in this chapter, you will find examples below of going between Celsius and Fahrenheit in two ways: quick-and-dirty and precise.

Celsius to Fahrenheit—Quick-and-Dirty

PROBLEM

You're in a lovely little hotel nestled in the Alps. You are all set for a wonderful hike, except you're wondering what the weather will be like. You look at the local paper, and it says that the day will be sunny, with a high of 18 degrees. What does that mean?

■ STEP 1

Let's start by doing it the easy way, to get a pretty close approximation. First, double (multiply by 2) the Celsius temperature (18).

18 times 2 = 36

■ STEP 2

Add 30 to the results of Step 1 (36).

36 plus 30 = 66

■ SOLUTION

The temperature is approximately 66 degrees Fahrenheit.

This approximation technique will overestimate Fahrenheit temperatures a little, particularly when temperatures get high. If it's really hot out, it's best to subtract a few degrees for a closer reading.

Celsius to Fahrenheit—Precise

You decide to use your calculator to get a more precise read on the weather.

■ STEP 1

Multiply the Celsius temperature (18) by 1.8.

18 times 1.8 = 32.4

■ STEP 2

Add 32 to the results of Step 1 (32.4).

32 plus 32.4 = 64.4

■ SOLUTION

The temperature in Fahrenheit is 64.4 degrees.

Fahrenheit to Celsius—Quick-and-Dirty

PROBLEM

Your fancy new car has many fun gadgets, including an outside temperature gauge. You decide to take your new car down to Mexico to see some old friends. As you're driving around, one of your friends comments on how hot it is. Your car's temperature gauge says that it's 84 degrees Fahrenheit. You want to tell your friends the temperature, but they are used to the Celsius scale. What temperature is it in Celsius?

STEP 1

Subtract 30 degrees from the Fahrenheit temperature (84).

84 minus 30 = 54

STEP 2

Divide the result of Step 1 (54) by 2.

54 divided by 2 = 27

SOLUTION

The temperature is approximately 27 degrees Celsius.

Fahrenheit to Celsius—Precise

PROBLEM

One of your friends is the skeptical type and takes out his calculator to check your estimate.

STEP 1

Subtract 32 degrees from the Fahrenheit temperature (84).

84 minus 32 = 52

STEP 2

Multiply the results of Step 2 (52) by .56.

52 times .56 = 29.12

SOLUTION

The precise temperature is 29.12 degrees Celsius.

(Tell your friend to give you a break. You were pretty close.)

In a Nutshell

TEMPERATURE

Celsius to Fahrenheit—Quick-and-Dirty

Fahrenheit temperature = (Celsius temperature × 2) + 30 degrees

Changing Celsius to Fahrenheit—Precise

Fahrenheit temperature = (Celsius temperature × 1.8) + 32 degrees

Fahrenheit to Celsius—Quick-and-Dirty

1. Subtract 30 degrees from the Fahrenheit temperature.

2. Celsius temperature = result of #1 ÷ 2

Fahrenheit to Celsius—Precise

Celsius temperature = (Fahrenheit temperature − 32) × .56

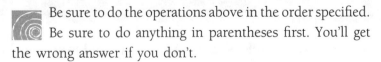 Be sure to do the operations above in the order specified. Be sure to do anything in parentheses first. You'll get the wrong answer if you don't.

Changing Currency

On the face of it, changing currency back and forth can seem very complicated. It is actually pretty straightforward, however, if you understand one basic concept—the *exchange rate*. The *exchange rate* measures how much money of one kind you can get for another kind. You can find exchange rates in most banks and in the business sections of major newspapers. They are most often listed two ways: "foreign currency in dollars" and "dollar in foreign currency." In other words, the exchange rate lists either how much one unit of foreign currency is worth in dollar terms or how much foreign currency one dollar will buy you.

Many Internet sites offer an easy way to determine exchange rates. You just plug in the currency you are exchanging "to" and "from," and they do the math for you.

Banks and exchange offices will exchange U.S. dollars to foreign currency based on their exchange rates. Built into their exchange rates is a fee for the transaction. (Beware, the exchange rate for buying foreign currency and selling the same foreign currency are often not the same.) Often, you can use your ATM card in a foreign country to get foreign currency. Your bank will then deduct from your balance the equivalent amount of dollars based on the appropriate exchange rate for that currency.

Exchange rates are generally pretty comparable between reputable financial institutions. However, even a small difference is important, particularly if you are exchanging a lot of money. Remember that you are looking for the exchange rate that gives you the highest amount of foreign currency for one dollar.

If you would rather not bother with exchanging money, some countries welcome you to spend U.S. dollars directly. Beware, though—the exchange rate in those situations usually isn't very good.

Changing money has recently gotten easier in much of Europe with the introduction of the euro (EUR). The euro is the standard piece of money now used in most of Europe, including France,

Spain, Italy, Ireland, Greece, Austria, Germany, and the Netherlands. Once you've exchanged your dollars for euros, you can travel most anywhere in Europe without having to change money as you go from country to country.

However, some countries, like Great Britain, have retained their own monetary systems, and there are close to fifty other foreign currencies listed in the business section of most major newspapers. To confuse things further, Canada has its own "dollar," which is worth considerably less than the U.S. dollar.

The most important thing to know about foreign currency is how to use it wisely in a foreign country, particularly if you are dealing with situations in which prices are not fixed. There are two ways of dealing with this challenge. The first is to know the exact exchange rate in each country in which you are traveling. You would then use your calculator to figure out how much an item or service in foreign currency would cost in dollars to see if the price is within your means. The second way is to form an approximate multiplier that would allow you (generally without your calculator) to easily estimate whether an item or service is reasonably priced. Below you will find examples of how to use the exact exchange rate and how to use an approximate multiplier.

You will also see an example of how to change currency when the exchange rate is very low. A very low exchange rate means that one unit of foreign money is worth only a small amount of American money, often in the tenths of a cent. For example, the exchange rate for a Japanese yen might be $.0075. You will therefore get a lot of foreign money for a small amount of dollars, and the prices in foreign currency will be in very big numbers, making conversion a little tricky.

Regardless of whether the exchange rate is high or low, it is important to know what it is for each currency you will be using and keep it handy at all times. Be sure to know the exchange rate in the form of what *one unit of foreign currency* equals in dollar terms.

Exchange rates can change dramatically over time. The exchange rates included in this book were accurate at the time it was written. Be sure to consult a newspaper, bank, or Internet site for up-to-date exchange rates.

Changing Currency—Precise Method

Let's start by heading off to Paris, where we will go through an example of how to use a precise method of exchanging currency.

PROBLEM

You are at a charming little shop in Paris, and you see a beautiful dress you are dying to have. The price is 225 euros. You have jotted down the exchange rate: 1 EUR = $.88. Can you afford it?

STEP 1

Multiply the price of the item in foreign currency (225 EUR) by the dollar value of one unit of foreign currency ($.88).

225 times $.88 = $198

SOLUTION

The dress costs $198.

It's a good idea to know the exchange rate *before* you travel so you can make advance hotel and other reservations wisely.

Changing Currency—Using an Easy Multiplier

Often, when traveling, you will forget where you've written down the exchange rate, or you just won't want to pull out your calculator every time you make a transaction. That's why it's a good idea to choose a multiplier that will approximate the exchange rate. Just pick a multiplier you can handle in your head or by jotting down a few numbers on a scrap of paper. As long as it is close to the exchange rate and easy to use, it does not matter exactly what it is. It is only meant as an approximation. Remember, though, if your multiplier is higher than the actual exchange rate, you will overestimate the cost. If it is lower, you will underestimate the cost.

PROBLEM

You and your husband are visiting your sister in London. You check the exchange rate in the paper before you go and see that one pound equals $1.43. You don't want to bother with your calculator and just want to have a rough idea of how much things cost. What multiplier should you use?

STEP 1

Round the exchange rate (how many dollars you get for *one* pound) to a number that's easy to multiply by.

Round 1.43 to 1.5, since it's easy to multiply by 1.5.

SOLUTION

Use 1.5 as a multiplier.

Now, let's see how to apply the multiplier you just picked.

PROBLEM

You have taken your sister out to dinner in London at a lovely restaurant, and the tab comes to 62 pounds. Approximately how much is that in dollars?

STEP 1

Round the tab (62 pounds) to the nearest multiple of ten, or tens place, so you can multiply it more easily.

The tens place is the second digit to the left. If a number is in the form "wxyz," z is the ones place, y is the tens place, x is the hundreds place, and w is the thousands place.

Round 62 to 60.

In rounding, you need to look at the number to the right of the place to which you are rounding. For example, when rounding 62 to the tens place, you look at the 2, since it is directly to the right of the tens place. If the number to the right is 5 or above, round up to the next multiple of 10. If it is less than 5, round down to the same "10." Since 2 is less than 5, you round 62 down to 60.

STEP 2

Multiply the tab from Step 1 (60) by your multiplier.

$60 times 1.5 = $90

1.5 makes a great multiplier, since it's easy to multiply by. Just take 1 of something, then add ½ of that to get the total. In the above example, 60 times 1.5 = 60 plus 30 (½ of 60) = 90.

SOLUTION

The tab is approximately $90.

You will often find yourself with a multiplier that is less than 1 and, therefore, in decimal form. Don't let that throw you. Just use the multiplier exactly as you did in the above example. For example, the price of a tour of Lima, Peru, might be sixty nuevo soles. Since one nuevo sol equals $.289, you choose a multiplier of .3. Simply multiply the price, sixty nuevo soles, by the multiplier *without* the decimal, which is 3 (60 times 3 = 180). Now just move the decimal over one place to the left (180. becomes 18.0), since your multiplier has one decimal place. (If your multiplier had two decimal places, you would have moved the decimal over two places.) The approximate price for the tour of Lima is $18.

Also, if your exchange rate is near to $.50, you can use .5 as a multiplier and just divide the foreign price by 2 (since $.50 is ½ of a dollar). Likewise, if your exchange rate is near to $.25, you can use .25 as a multiplier and just divide the foreign price by 4 (since $.25 is ¼ of a dollar).

Don't get too hung up on using an exact multiplier. All you're trying to do is get a rough idea of what something's costing you. Use a number you can work with easily in your head.

Changing Currency When Exchange Rates Are Very Low

As discussed above, very low exchange rates are sometimes tough to handle. Don't despair. You can use the same two approaches you used above. You just need to make some adjustments if you prefer to use an approximate multiplier. The example below will first show you how to use a precise method when the exchange rates are very low. Then it will explain how you might handle the same problem with a multiplier.

PROBLEM

You are traveling to Japan and are making hotel reservations for a stay in Tokyo. The reservations agent says that the room goes for 25,400 yen a night. Wow. You know things are expensive there, but that sounds outrageous. Is it? The exchange rate is one yen = $.0076.

Precise Method

STEP 1

Multiply the price in yen (25,400 yen) by the exchange rate ($.0076)

25,400 yen times $.0076 = $193

SOLUTION

The room will cost $193 per night.

Easy Multiplier
Let's take a look at the same problem using a multiplier. When the exchange rate is very low, you need to put it in a form you can handle, by getting rid of the zeroes. Let's see how this works.

PROBLEM

STEP 1

You know your exchange rate: 1 yen = $.0076. Move the decimal point of the dollar equivalent *to the right* until you get a number in the ones place that is not 0. As you do this, count the number of places you move the decimal point.

.0076 becomes 7.6 in 3 moves of the decimal place.

 The ones place is the number directly to the left of the decimal.

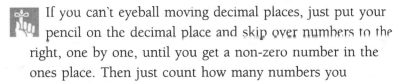 If you can't eyeball moving decimal places, just put your pencil on the decimal place and skip over numbers to the right, one by one, until you get a non-zero number in the ones place. Then just count how many numbers you skipped.

STEP 2

Round the multiplier from Step 1 (7.6) to a whole number (one with no decimals) so you can use it more easily.

multiplier = 8

STEP 3

Round the hotel price (25,400 yen) to a number you can manipulate easily in your head.

25,000 yen

■ STEP 4

Take the rounded price from Step 3 (25,000) and move the decimal *to the left* the number of moves from Step 1 (3).

25,000. moved 3 decimals to the left becomes 25.000, which equals 25.

 If there are only zeros after the decimal place, you can just ignore them. They have no value.

■ STEP 5

Multiply the result of Step 4 (25) by your multiplier from Step 2 (8).

25 times 8 = 200

■ SOLUTION

Your hotel room costs approximately $200.

(Compare your $200 estimate with your $193 precise conversion. You realize you are pretty close, so your multiplier is okay.)

Don't worry if finding a multiplier seems complicated. Remember, you only have to compute the multiplier once for each currency. Once you use it a couple of times, it will become second nature to you.

Canada is on a dollar system just like the United States, but the Canadian dollar is currently worth considerably less than the U.S. dollar. It's a little confusing to travel in Canada because everything is marked in dollars, but they are *Canadian* dollars. Use the techniques above to change the Canadian dollar to U.S. dollars. Just make sure you keep your units straight. You can use U.S. dollars in many Canadian stores, but be careful. Be sure the clerk converts your U.S. dollars to Canadian dollars, then takes out the purchase price.

In a Nutshell

CHANGING CURRENCY

Changing Currency—Precise Method

1. Get the exchange rate in this form: *one unit of foreign currency = so many U.S. dollars.*

2. U.S. dollars to buy something = the dollars part of the exchange rate (the dollar equivalent of one unit of foreign currency) × price in foreign currency.

Changing Currency—Using an Easy Multiplier

1. Get the exchange rate in the form: *one unit of foreign currency = so many U.S. dollars.*

2. Multiplier = the dollars part of the exchange rate rounded to a number you can work with easily in your head.

3. Round the price of the product or service to a number you can work with easily in your head.

4. U.S. dollars to buy something = multiplier (from #2) × the rounded price in foreign currency (from #3).

Changing Currency When Exchange Rates Are Very Low

Precise Method

Use the method as described above in "Changing Currency—Precise Method."

Easy Multiplier

1. Get the exchange rate in the form: *one unit of foreign currency = so many U.S. dollars.*

2. Keep moving the decimal place *to the right* until you get a number in the ones place that is *not* "zero." Round this number to a number you can work with easily in your head. This is your *multiplier.* Count the number of moves of the decimal you made.

continued

continued from previous page

3. Round the price of the product or service to a number you can work with easily in your head.
4. Take the rounded price in foreign currency (from #3) and move the decimal *to the left* the number of moves you made in #2.
5. U.S. dollars to buy something = multiplier (from #2) × the result of #4.

Whew. There is a lot of math in traveling. But now that you've gotten a handle on it, relax and enjoy yourself.

7 | Sports

Have you ever wondered what all those numbers in the sports pages mean? If you haven't, you can skip this chapter. If you have, then the first thing you need to know is that sports stats are really pretty simple, once you get past all the jargon and the abbreviations. Most of them are self-explanatory. "The field goal kicker was three for four today" means he made three out of the four field goals he attempted. "The goalkeeper had eight saves" means that she saved the ball or the puck from going into the goal eight times. "The batter had four RBIs (runs batted in)" means his hits resulted in four runs being scored.

Even the terms that are not self-explanatory are pretty straightforward. This chapter will explain how to compute and understand some of the most common sports statistics. You will learn the basics of how to read standings and how to use percentages to measure sports performance. Then some statistics for individual sports will be explained: earned run average and batting average in baseball; yards per carry, reception, or punt, as well as quarterback rating, in football; goals against average in hockey; and par and handicaps in golf.

Even if you're not that interested in sports, just think how impressive you will be around the office coffeemaker on Monday mornings.

Standings

A standings table is used in most sports to let you know where your team is relative to the other teams in its conference or league. The teams are listed in the order of their ranking. Most sports have a standings chart that looks something like the following:

Standings: National Tiddlywink League				
Team	W	L	Pct.	GB
Podunk	7	4	.636	—
East Suburbia	6	5	.545	1
West End	5	6	.455	2
Metropolis	4	7	.364	3

Games Won/Lost

The teams are listed in the order of their standings in the league, with Podunk on top right now. The "W" column lists the number of games the team has won. The "L" column lists the number of games the team has lost. Easy enough so far.

Percentage of Games Won

While there are no "%" signs, the "Pct." column does give you the percentage of its games an individual team has won. For reasons that are not entirely clear, the "Pct." column gives you the percentage in decimal form. A "1.000" in this column means the team has a perfect record and won all, or 100 percent, of their games. A ".500" means the team has won half, or 50 percent, of their games. In the above table, Podunk has won 63.6 percent of its games, East Suburbia 54.5 percent, West End 45.5 percent, and Metropolis only 36.4 percent.

 To change a decimal into a percent, just move the decimal point to the right two places and add a "%" sign.

"Percentage of games won" is calculated by dividing the games won by a particular team by the total games it has played. For example, the East Suburbia Eagles have played 11 games (6 won plus 5 lost = 11). Divide the 6 games it has won by the 11 it has played to get a "Pct." of .545.

Games behind the Leader

"GB" means *games back*, or number of games behind the leader. It really means how many games your team would have to win and the leader would have to lose for the teams to be tied for first place. To calculate the number of games behind the leader your team is, add the difference in the number of wins between your team and the leader to the difference in the number of losses, and divide the total by 2. For example, to calculate games back for the East Suburbia Eagles, you see that the difference in the number of wins between it and the leading Podunk Patriots is 1 (7 minus 6), and the difference in the number of losses is also 1 (7 minus 6). Add the numbers together to get 2 (1 plus 1) and divide that total (2) by 2 to get the games-back statistic, "1" (2 divided by 2 = 1).

Because of scheduling, teams will often have played different numbers of games, and the "GB," or "Games Back," column could include ⅓s.

In a Nutshell

STANDINGS

W = number of games the team has won

L = number of games the team has lost

Pct. = number of games won ÷ number of games played

GB = {(wins by team in first place − wins by this team) + (losses by this team − losses by team in first place)} ÷ 2

NOTE: If the difference in wins or the difference in losses is a negative number (e.g., 5 − 8 = −3) turn it into a positive number (e.g., −3 becomes 3).

Do the steps inside the parentheses first, the steps inside the brackets next, and then finish any other operations.

Using Percentages to Measure Sports Performance

Many sports statistics include percentages as a way of measuring performance. In basketball, you might hear the announcer talking about field goal percentage or free throw percentage. In football, a quarterback is often judged on his pass completion percentage, interception percentage, or touchdown percentage.

Remember that percentages are just a way of getting at what portion one thing is of another. Twenty-five percent is one-fourth of something; 50 percent is half of something; 75 percent is three-quarters of something; 100 percent is all of something. If a basketball player is "shooting 75% from the free throw line," she is making three-quarters of the free throws she tries. If a quarterback is "passing at about 50%," he is completing one-half of his passes.

Look at the following example to see how you would compute a percentage in order to measure sports performance.

 For more review on percents, see appendix 3.

PROBLEM

Your daughter has become a hot player for her high school basketball team. You hear that a college scout will be in the stands for the big game tomorrow night, and you would like to share some of your daughter's statistics with her. You have been carefully tracking her scoring throughout the season. She has attempted 127 shots (not including foul shots) and has made 91 of them. What is her shooting, or field goal, percentage?

STEP 1

Divide the number of shots your daughter made (91) by the number of shots she attempted (127).

91 divided by 127 = .72

STEP 2

Multiply the result of Step 1 (.72) by 100.

.72 times 100 = 72

SOLUTION

Your daughter's field goal percentage is 72%.

Any sports percentage can be calculated the same way. Just divide the number the player got right by the total she attempted. Then multiply that result by 100, and you have a percent. For example, a quarterback's completion percentage is just the number of passes the quarterback made divided by the number of passes attempted then multiplied by 100.

Some football percentages can become a little confusing. For example, a quarterback's touchdown percentage is calculated by dividing the number of touchdown passes thrown by the *total* number of passes (not just touchdown passes) attempted.

In a Nutshell

USING PERCENTAGES TO MEASURE SPORTS PERFORMANCE

Percentage = (the number the player got right ÷ the total number attempted) × 100

 Be sure to do the step inside the parentheses first, then multiply by 100.

Sport-Specific Stats

The statistics discussed above can be found in a number of different sports. There are a number of statistics, however, that are specific to one or two sports. They usually have their own abbreviations and jargon. There are a number of books on sports that will define these terms more completely. Included below are examples of how to compute a few of the more familiar ones.

Baseball

Earned Run Average

Earned run average (ERA) is a measure of pitching proficiency. It measures the average number of runs a pitcher allows in nine innings of play. The smaller the ERA, the better. The ERA includes a mechanism to allow for a pitcher who pitches less than a full nine-inning game. To calculate ERA, just multiply the number of runs the pitcher has allowed by 9 and then divide that number by the number of innings he has pitched.

PROBLEM

Your son's baseball coach tells you that your son has an ERA of 2.29, and he might be the next Cy Young. While you dutifully go to each and every game, you don't know a lot about baseball. You thought ERA was the Equal Rights Amendment and don't have a clue who Cy Young is. Your son has pitched in 55 innings and has allowed 14 runs. How did his coach calculate his ERA?

STEP 1

Multiply the number of runs your son has allowed (14) by 9.

14 times 9 = 126

STEP 2

Divide the result of Step 1 (126) by the number of innings he has pitched (55).

126 divided by 55 = 2.29

SOLUTION

That's how the coach calculated your son's ERA to be 2.29.

(By the way, Cy Young was a very famous pitcher who played at the turn of the last century.)

An earned run average is based only on the "earned" runs a pitcher allows. If a player who scores arrives on base through a fielding error or by the pitching of a previous pitcher, that run is "unearned" and is not counted against the current pitcher.

Batting Average

The term *batting average* is somewhat misleading, because it doesn't really tell you outright how many hits a player makes, on average, in a game. Rather, the batting average is the percentage of hits compared to total times at-bat. And, to further complicate matters, this batting percentage is in decimal form. To calculate batting average, simply divide the total number of hits a player has by the number of times she has been at bat.

As is the case with much sports talk, since the term *batting average* has been used for years, sports fans are familiar with its usage. If a ballplayer "hits 350," most sports fans know that the player is having a great year, even though they might not know what it actually means. (By the way, it means the batter hits about 35% of the times he is at bat.)

PROBLEM

You have joined an amateur softball league for the season. You are hitting well and are surprised when the coach

tells you your batting average is only .244. You want to double-check, so you look at the stats and see that you have 45 at-bats and 11 hits. What is your batting average?

STEP 1

Divide the number of hits (11) by the number of at-bats (45).

11 divided by 45 = .244

SOLUTION

Your batting average is .244. Your coach was right.

(This is also called "batting 244.")

When computing batting average, do not include the following in the total number of at-bats: walks, sacrifice flies or bunts, or being hit by a pitch. None of these counts as a hit, either. However, if a hitter gets to base on an error, he or she is charged with an at-bat but not given credit for a hit.

Football

Yards per Carry, per Reception, per Punt

In football, there are a number of terms that are designed to give an idea of how many yards, on average, a player advances the ball. A running back might have a "yards per carry" stat; a punter, "yards per punt;" a receiver, "yards per reception." They are all calculated the same way. Just divide the total number of yards per carry, punt, or reception by the number of times the player carried, punted, or received the ball.

PROBLEM

Your favorite pro running back, Ricky "The Rabbit" Rogers, has carried the football 45 times for 567 yards this season. What is his stat for average yards per carry?

STEP 1

Divide the number of yards he has carried the ball (567) by the number of times he has carried it (45).

567 divided by 45 = 12.6

SOLUTION

The Rabbit is averaging 12.6 yards per carry.

Many sports averages can be handled in the same way. Just take the amount of ground a player covers and divide it by the number of times he or she carries, passes, receives, or kicks the ball.

Quarterback Rating

Because successful quarterbacks must demonstrate a number of skills, a quarterback rating system was designed some years ago. It combines four factors in a complicated formula to come up with a total number of points. The higher the number of points, the better the quarterback. A perfect score is 158. Most sports fans, even the most rabid, do not know how to calculate a quarterback rating. But if you really want to show off to your sports minded friends, just carefully follow the steps below.

PROBLEM

The quarterback for your local professional football team is having a great year. So far, he has made 45 out of 60 passes for a total of 525 yards. He has thrown 4 touchdown passes and has only been intercepted once. What's his quarterback rating?

STEP 1

Divide the number of passes made (45) by the number of passes attempted (60).

45 divided by 60 = .75

STEP 2

Subtract .3 from the result of Step 1 (.75).

.75 minus .3 = .45

STEP 3

Divide the result of Step 2 (.45) by .2.

.45 divided by .2 = 2.25

(If this number is less than 0, substitute 0. If this number is greater than 2.375, then substitute 2.375.)

STEP 4

Divide the total number of yards completed (525) by the number of passes attempted (60).

525 divided by 60 = 8.75

STEP 5

Subtract 3 from the result of Step 4 (8.75).

8.75 minus 3 = 5.75

STEP 6

Divide the result of Step 5 (5.75) by 4.

5.75 divided by 4 = 1.44

(If this number is less than 0, substitute 0. If this number is greater than 2.375, then substitute 2.375.)

STEP 7

Divide the number of touchdown passes (4) by the number of passes attempted (60).

4 divided by 60 = .07

STEP 8

Divide the result of Step 7 (.07) by .05.

.07 divided by .05 = 1.4

(If this number is less than 0, substitute 0. If this number is greater than 2.375, then substitute 2.375.)

STEP 9

Divide the number of interceptions (1) by the number of pass attempts.

1 divided by 60 = .02

STEP 10

Subtract the result of Step 9 (.02) from .095.

.095 minus .02 = .075

STEP 11

Divide the result of Step 10 (.075) by .04.

.075 divided by .04 = 1.88

(If this number is less than 0, substitute 0. If this number is greater than 2.375, then substitute 2.375.)

STEP 12

Add the results of Step 3 (2.25), Step 6 (1.44), Step 8 (1.4) and Step 11 (1.88).

2.25 plus 1.44 plus 1.4 plus 1.88 = 6.97

STEP 13

Multiply the result of Step 12 (6.97) by 100.

6.97 times 100 = 697

STEP 14

Divide the result of Step 13 (697) by 6.

697 divided by 6 = 116

SOLUTION

The quarterback rating for your quarterback is 116.

(That's terrific by the way. Breaking 100 is great.)

Hockey

Goals-Against Average

Hockey uses a statistic to evaluate a goalie's performance called *goals-against average,* or GAA. The goals-against average is similar to the earned run average for baseball pitchers. It measures the average number of goals that were scored against a goalie in a sixty-minute game. The goals-against calculation allows for the fact that a goalie might not play for the entire sixty minutes of the game. To calculate GAA, multiply the number of goals scored against a goalie by 60, then divide that number by the total number of minutes he has played.

Look at the following example.

PROBLEM

Your sister is a goalie for an amateur ice hockey team. She has played 650 minutes so far this season, and she has allowed 25 goals. What is her goals-against stat so far this season?

STEP 1

Multiply the number of goals scored (25) by 60.

25 times 60 = 1,500

STEP 2

Divide the result from Step 1 (1,500) by the total minutes played (650).

1,500 divided by 650 = 2.30

SOLUTION

Your sister's goals-against average for the season is 2.30.

Golf

The basic idea in golf is to use the fewest shots possible to get the ball from the tee into the hole. You generally play nine or eighteen "holes." Keeping score is, therefore, just a matter of adding up the scores for individual holes. Unlike in other sports, in golf the lowest score wins.

Par

Golfers use "par" to indicate the number of shots it should take you, on a given hole and golf course, to get your ball from the tee into the hole. If the golf course gurus decide that a hole has a par 3, then it should take a golfer three shots to get the ball from the tee into that hole. Par for each hole is added together to get a total par for the course. You can then compare your total score to the total par for the course. If a golfer is −7 for the round, that means that she shot seven strokes below par for the whole course. If the golfer is +5, he shot five strokes above par for the whole course.

Take a look at the following example to see how to measure performance relative to par.

PROBLEM

You are a golf fanatic and are thrilled to be able to attend your first big golf tournament. You are following your

favorite golfer, Patty "The Putter" Pastorelli. On the first five holes, she has scored the following:

Hole	Par	Score
1	3	3
2	4	5
3	3	2
4	3	3
5	4	3

How's Patty doing relative to par?

STEP 1

Total Patty's scores and the pars for the holes.

Hole	Par	Score
1	3	3
2	4	5
3	3	2
4	3	3
5	4	3
Total	17	16

STEP 2

Subtract total par (17) from Patty's total (16).

16–17 = –1

Don't panic when you see negative numbers. All a "–" means in front of a number in golf is that Patty shot that many strokes *less than* par.

SOLUTION

So far, Patty is –1, or 1 under par, for the round.

Handicap

Golf uses a handicap system to allow players of varying ability to play with each other competitively. A handicap gives you an idea of how you play, on average, relative to par. If you have an 8-handicap, you should generally shoot eight strokes over par. At the end of a round of golf, you total the number of shots for each hole, then subtract the amount of your handicap to give you your final score.

Calculating your handicap can get complicated and involves knowing about things like course ratings and slopes. The best way to get your handicap is to keep track of your recent scores and then ask a professional at the club or the course at which you play to calculate your handicap for you. Computer software and Internet sites that do the calculation for you are available as well.

The following example will show you how to compute your score with a handicap once it is computed.

PROBLEM

You have just completed an eighteen-hole round of golf with your brother, who is an excellent golfer and has a 5-handicap on your local course. Your handicap is 19. His total score was 78. Your score was 90. Who won the round?

STEP 1

Subtract your brother's handicap (5) from his shot total (78) to get his final score.

78 minus 5 = 73

STEP 2

Subtract your handicap (19) from your shot total (90) to get your final score.

90 minus 19 = 71

SOLUTION

Your brother scored 73; you scored 71. You won.

In a Nutshell

SPORT-SPECIFIC STATS
Baseball
earned run average (ERA) = (earned runs × 9) ÷ innings played by the pitcher

> earned run = a run that is the pitcher's responsibility. It does *not* include runs scored through fielding errors or by the pitching of a previous pitcher.

batting average = number of hits by an individual player ÷ number of at-bats

> at-bats = the number of times a player went to the plate to bat, excluding walks, sacrifice flies or bunts, or being hit by a pitch.

Football
yards per carry/reception/punt = total yards a player carried/received/punted the ball ÷ total number of carries/receptions/punts

Passer Rating

1. Calculate: (pass completions ÷ pass attempts) −.3

2. Divide the result of #1 by .2

3. Calculate: (total yards passed ÷ pass attempts) −3

4. Divide the result of #3 by 4

5. Calculate: (touchdown passes ÷ pass attempts)

6. Divide the result of #5 by .05

7. Calculate: .095 − (interceptions ÷ pass attempts)

continued

continued from previous page

8. Divide the result of #7 by .04

9. If any of the calculations in Steps 2, 4, 6, or 8 results in a number less than 0, then substitute 0 for that answer.

10. If any of the calculations in Steps 2, 4, 6, or 8 results in a number greater than 2.375, then substitute 2.375 for that answer.

11. Add the results of #2, #4, #6, and #8.

12. quarterback rating = (results of #11 × 100) ÷ 6

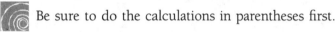 Be sure to do the calculations in parentheses first.

Hockey

goals-against average (GAA) = (goals scored against a goalie × 60) ÷ total number of minutes played by the goalie

 Do the calcultion inside the parenthesis first, then do the division.

Golf

Handicap

final score with handicap – total number of strokes in a round – handicap

By now, you should be pretty impressed with your ability to sling around the old sports stats. You might not be the next Babe Ruth or Martina Navritalova, but you will sure talk a good game.

3 Personal Finance Math

8 | Wages, Salaries, and Taxes

You probably remember when you got that first big job of yours flipping burgers at the Burger Barn. You carefully multiplied your wage of two dollars per hour by the forty hours a week you were working that summer and figured you would be making a cool $80 a week. You would have that new car in no time. Then you got your first paycheck. Where did all your money go? Suddenly, you remember your father quoting some guy who said something about death and taxes.

Answering the question of where your money goes is often no easy matter. Few things are as confusing as the federal tax code. And have you ever noticed that every time Congress passes a "tax simplification act" taxes get more complicated? To help me write this chapter, a tax professional gave me a copy of a federal tax handbook that is supposed to offer a clear and easy-to-understand explanation of the tax system. It is 779 pages long.

Let's take a deep breath and start at the beginning. For most of us, salaries and taxes are actually pretty straightforward. This chapter begins with an explanation of how your wages are calculated and how salary increases and decreases are figured. It goes on to explain all those items that come out of your paycheck every week. We will also plunge right into some tax basics, like how tax brackets break down, how to use tax tables and tax formulas, and whether you should itemize deductions or take the standard deduction.

Wages

Often, workers are paid an hourly wage to do a certain job. To figure out your paycheck (before deductions), you just multiply the hours worked that week by your hourly wage. If you work forty hours in a given week, and you make $10 an hour, your paycheck will be $400 (40 times $10) before deductions.

Time and a Half and Double Time

Under special circumstances, like holidays and weekends, often the hourly wage is increased to time and a half or double time. *Time and a half* is just payment of 1.5 times your normal hourly wage; *double time* is twice what you normally earn per hour.

PROBLEM

You are saving up to take your husband on a Caribbean cruise. You have worked 40 hours this week at your normal hourly pay of $16.50. You have taken in another 8 hours of overtime at time-and-a-half pay. In addition, you plan to work 4 hours of double time on the upcoming holiday. How much should your weekly paycheck be before taxes?

STEP 1

Calculate the amount of regular pay you will receive by multiplying the number of normal hours worked (40) by your normal hourly pay ($16.50).

40 times $16.50 = $660.00

STEP 2

Compute your hourly time-and-a-half overtime pay by multiplying your normal hourly pay ($16.50) by 1.5

$16.50 times 1.5 = $24.75

■ STEP 3

Calculate the amount of time-and-a-half overtime pay you will receive by multiplying the number of time-and-a-half hours worked (8) by the hourly time-and-a-half pay from Step 2 ($24.75).

8 times $24.75 = $198.00

■ STEP 4

Compute your hourly double-time pay by multiplying your normal hourly pay ($16.50) by 2.

$16.50 times 2 = $33.00

■ STEP 5

Calculate the amount of double-time pay you will receive by multiplying the number of double-time hours worked (4) by the double-time hourly pay from Step 4 ($33.00).

4 times $33.00 = $132.00

■ STEP 6

To compute your total weekly pay, add your normal pay from Step 1 ($660.00), your time-and-a-half pay from Step 3 ($198.00), and your double-time pay from Step 5 ($132.00).

$660.00 plus $198.00 plus $132.00 = $990.00

■ SOLUTION

Your total weekly pay before taxes should be $990.00.

(At this rate, you will be in the Caribbean in no time.)

Salaries—Percent Raises

Those who are not paid hourly are usually paid a yearly salary, which does not vary depending on the hours worked.

Calculating a Raise When You Know the Percentage Increase

Salaried employees often get raises based on a certain percentage. In particular, cost-of-living increases are usually expressed with a percentage. To compute the actual salary increase, simply divide the percentage increase by 100, and multiply it by your salary. For example, someone making $50,000 a year and getting a 3 percent cost-of-living increase would receive a $1,500 increase (3% =.03, and .03 times $50,000 = $1,500). (For more help with percents, see appendix 3.) Just add that increase to your current salary to get your new salary.

PROBLEM

Your boss has given you an excellent rating for your yearly evaluation. As a result, you will be getting a 14% raise. You currently make $38,000 a year. How much will you be making with the raise?

STEP 1

For ease in handling, change your percentage raise (14%) into a decimal by dividing the percentage by 100.

A percentage is just a number divided by 100. An easy way to convert a percentage to a decimal value is to move the decimal place of the percentage to the left two places. If there is no decimal point, just put one at the end of the number. If you have trouble visualizing moving decimal places, just put your pencil point on the decimal point and move it two digits to the left.

14 divided by 100 = .14

STEP 2

Calculate your actual raise by multiplying the percentage of increase in decimal form from Step 1 (.14) by your salary ($38,000).

.14 times $38,000 = $5,320

STEP 3

To calculate your new salary, add your increase from Step 2 ($5,320) to your old salary ($38,000).

$5,320 plus $38,000 = $43,320

SOLUTION

Your new salary will be $43,320.

Calculating Percentage Increase

A salaried person will often be given a specific raise without any information on how big a percentage the increase is. To find the percentage increase, just subtract your old salary from your new salary, then divide the difference into your old salary. Multiply that number by 100.

PROBLEM

Your company has just raised your salary from $48,000 to $52,500. You are curious how your raise fares against the 10% average annual company raise. By what percentage did your salary increase?

STEP 1

Compute the amount of your raise by subtracting your old salary ($48,000) from your new salary ($52,500).

$52,500 minus $48,000 = $4,500

STEP 2

Divide your raise from Step 1 ($4,500) by your old salary.

$4,500 divided by $48,000 = .094

When you are calculating either percentage increase or percentage decrease, remember always to divide by the *old* number, *not* the new one.

STEP 3

To get the decimal from Step 2 (.094) into percent form, just multiply it by 100.

.094 times 100 = 9.4%

To get a decimal value into percentage form, just multiply it by 100. An easy way to multiply a number by 100 is to move the decimal point two places to the right.

SOLUTION

Your raise is 9.4%, a little below the company average.

Calculating Percentage Decrease

Sometimes you will be faced with situations where your salary actually decreases, and you might want to calculate what the percentage of decrease is. You compute percentage decrease very much like you calculated percentage increase. Find the difference between your old and new salary by subtracting your old salary from your new salary. Then divide this figure by your old salary and multiply by 100 to find the percentage decrease.

PROBLEM

You have given up your lucrative corporate job to follow your life's passion of studying the endangered multi-

winged mottled moth. Your salary has decreased from $65,000 to $35,000. By what percentage did your salary decrease?

STEP 1

Calculate the amount of your salary decrease by subtracting your new salary ($35,000) from your old salary ($65,000).

$65,000 minus $35,000 = $30,000

STEP 2

Divide your salary decrease from Step 1 ($30,000) by your old salary ($65,000).

$30,000 divided by $65,000 = .46

STEP 3

Multiply the result of Step 2 (.46) by 100 to get percentage decrease.

.46 times 100 = 46%

SOLUTION

Your salary decrease with your new career is 46%.

It's a large percentage decrease, but a small price to pay for personal fulfillment. (Not to mention that the moths will appreciate your efforts.)

The Rich Get Richer

Did you ever notice that no matter how many great raises you get, the people at the top just seem to be getting farther and farther ahead? It is not your imagination. Let's look at the numbers.

Look at the following chart:

	Beginning Salary	Year 1	Year 2	Year 3	Year 4
Your Colleague	$60,000 (10% raise)	$66,000	$72,600	$79,860	$87,846
You	$30,000 (15% raise)	$34,500	$39,675	$45,626	$52,470
Difference	$30,000	$31,500	$32,925	$34,234	$35,376

Let's say your starting salary is $30,000, and you work very hard and are rewarded with regular 15 percent raises for four years. Your colleague's starting salary is $60,000. He does average work and is rewarded with regular 10 percent raises for four years. Even though your work is better, and you have been rewarded with higher-percentage raises, your colleague's salary continues to grow faster than yours does. In fact, the difference between your two salaries at the end of the four years is over $5,000 more than when you started ($35,376 in year 4 minus $30,000 at the beginning.)

Because percentage raises favor employees with higher salaries, it is important to come into a job at the highest salary you can negotiate. And if you are stuck in a lower-paying job, you might need to make a fuss to get yourself a one-time jump in pay to even things out.

Why Is Your Paycheck So Small?— Payroll Deductions

Despite periodic raises or working extra hours at time and a half, it might seem that you are still not getting the amount of money at the end of the week that you anticipate. That is because there are a number of deductions that come out of your paycheck before you even see it.

Your pay stub might look something like the one below.

Earnings		this period	year to date
	Gross Pay	**1,507.00**	
Deductions	**Statutory**		
	Federal Income Tax	-173.67	1,053.36
	Social Security Tax	-88.91	536.05
	Medicare Tax	-20.80	125.37
	DC State Income Tax	-88.63	535.68
	Other		
	Dental-Classic	-6.00	36.00
	Health	-67.00	000.00
	Health Club	-25.00	125.00
	401-K	-60.28	361.68
	Net Pay	**$976.71**	

The number at the top should be your *gross* pay, or the pay you start with before anything is taken out. From your gross pay, several withdrawals (or "withholdings") are made, which most likely include the following:

- federal taxes
- state and local taxes
- FICA (that's social security and Medicare)

Withdrawals might also include:

- health insurance premiums
- contributions to cafeteria plans
- union dues
- 401(k) contributions
- United Way or similar contributions

Remember that the amount of federal and state taxes taken out of your salary is just a preliminary estimate of your taxes based on the information you provided in your W-2 form. When you file your taxes, you will deduct the amount paid from your regular paycheck from the total amount you owe. Therefore, it is best to get the amount of taxes taken out of your paycheck pretty close to the real amount of taxes you will owe on April 15. Otherwise, you might be in for a rude shock.

Your FICA contribution is 7.65 percent of your gross salary, of which 1.45 percent is Medicare. You always pay that amount regardless of your salary. The remaining 6.2 percent of the FICA goes to social security. For tax year 2002, you pay social security only on the first $84,900 you earn. This salary cap increases annually.

After everything is deducted, what you have left is *net* income, or that part of your paycheck that is yours to do with what you like. It might be a small comfort, but at least now you know where your money is going.

In a Nutshell

WAGES AND SALARIES
Wages

Time and a Half

time-and-a-half hourly wage = 1.5 × normal hourly wage

total time-and-a-half wages = hours at time and a half × time-and-a-half hourly wage

Double Time

double-time hourly wage = 2 × normal hourly wage

total double-time wages = hours at double time × double-time hourly wage

Salaries

Calculating a Raise and New Salary When You Know the Percentage Increase

percent in decimal form = percent ÷ 100

continued

continued from previous page

raise = percent in decimal form × old salary

new salary = old salary + raise

Calculating Percentage Increase

raise = new salary − old salary

percentage increase = (raise ÷ by old salary) × 100

Calculating Percentage Decrease

decrease = old salary − new salary

percentage decrease = (decrease ÷ old salary) × 100

 Make sure you do the steps in the parentheses first. Otherwise, you will get the wrong answer.

Taxes

The biggest payroll deduction most of us have is, of course, taxes. You probably know that the proportion of your salary that you pay in taxes goes up as your salary goes up. Under current tax law, most people pay between 10 percent and 38.6 percent of their salaries in federal taxes. (By the way, the higher rates are slated to go down over the next few years.) State and local taxes vary from state to state but usually take a much smaller bite out of your salary.

What you might not have realized, however, is that different parts of your salary are taxed at different rates. The first part of your salary that you earn is always taxed at a relatively low rate regardless of how high your salary is. Even if you are married and you and your spouse make $500,000 a year, the first $12,000 is always taxed at 10 percent. The next parts of your salary are taxed at increasing rates until you get to the top rate of 38.6 percent. This phenomenon is reflected in the tax tables in the instruction book that accompanies your annual tax form from the IRS.

Remember that you are only taxed on the portion of your income that is subject to taxes, or *taxable income*. Taxable income will differ between individuals, and your tax form will walk you through the calculation. To get from gross income (total income) to taxable income, you usually subtract such things as exemptions for yourself and your children, any deductions you are allowed, and contributions to certain retirement funds.

All the tax calculations in this chapter are based on 2002 rules and regulations and are for illustration only. Up-to-date tables, formulas, deductions, and other tax information are in the instructions that accompany your tax form.

Using Tax Tables

To determine your federal income tax if your taxable income is less than $100,000, just look up your taxable income in the tax tables that come with your tax form. The IRS has done the math for you. If your taxable income is $100,000 or more, you will need to use the formula in the instruction book to determine your tax. Find the table for your filing status and follow the directions.

Look at the tax formula table below for married couples filing jointly:

2002 Federal Income Tax for Married Couples Filing Jointly

If taxable income is:	The tax is:
Not over $12,000	10% of taxable income
Over $12,000 but not over $46,700	$1,200.00 plus 15% of the excess over $12,000
Over $46,700 but not over $112,850	$6,405.00 plus 27% of the excess over $46,700
Over $112,850 but not over $171,950	*$24,265.50 plus 30% of the excess over $112,850*
Over $171,950 but not over $307,050	$41,995.50 plus 35% of the excess over $171,950
Over $307,050	$89,280.50 plus 38.6% of the excess over $307,050

Let's walk through an example of how to use this table for a couple who is married and filing jointly and earned $135,650.

PROBLEM

You are determined to do your taxes yourself this year. You and your husband earned $135,650 for 2002 and are filing jointly. You have $12,530 in itemized deductions and take two personal exemptions totaling $6,000. You subtract these amounts from your gross income of $135,650 to come up with taxable income of $117,120. How much federal income tax do you owe?

STEP 1

Since your taxable income is above $100,000, you need to turn to the tax rate table in the tax form instructions. Find your taxable income ($117,120) on the tax rate table (page 168).

STEP 2

Determine "the excess over $112,850" by subtracting $112,850 from your taxable income ($117,120).

$117,120 minus $112,850 = $4,270

STEP 3

Find "30% of the excess" by multiplying the excess determined in Step 2 ($4,270) by 30%. First convert 30% to a decimal for ease in handling.

30% = 30 divided by 100 = .30

An easy way to divide a number by 100 is to move the number's decimal two places to the left. If there is no decimal, just put one at the end of the number. If you have trouble imagining moving decimals, just put your pencil point on the decimal point and move it two numbers to the left.

Then multiply the excess from Step 2 ($4,270) by the decimal form of 30%.

.30 times $4,270 = $1,281

■ STEP 4

To calculate your final tax, the table tells you to add $24,265.50 to the result of Step 3 ($1,281).

$1,281 plus $24,265.50 = $25,546.50

■ SOLUTION

Your federal income tax is $25,546.50.

(That wasn't too bad, was it? Well, paying the tax might be.)

Can a Raise Be Bad for You?

You might have heard a colleague say, "That raise ended up costing me money because it put my whole salary in the next tax bracket." Now that you understand taxes a little, you realize that phenomenon is simply impossible. The lower rates you enjoy on the first parts of your salary never change, regardless of how big your income gets. A $5,000 raise might throw the last part of your earnings into a new tax bracket but never puts your entire taxable income into a new tax bracket.

Your state taxes work much the same way as federal taxes. Most states have similar tables or formulas, but the tax rates are considerably lower.

Certain individuals with modest incomes might be eligible for an "earned income tax credit," which not only exempts them from having to pay taxes, but also actually pays them some extra money in the form of a tax credit. Check with the tax instruction book to see if you qualify.

Should You Itemize Your Deductions?

As noted before, you get a break on your taxes by not having to pay taxes on certain items. You can "deduct" from your taxable income things like state and local taxes, interest on your house, property tax on your house, and medical expenses if they get extreme. These deductions are all detailed in the IRS instructions accompanying your tax form.

You have two choices when it comes to these deductions from your taxable income. You can keep track of all the individual deductions that are allowed and "itemize," or list, them separately on your return. Or you can skip the hassle and take the "standard deduction." The standard deduction for various types of taxpayers for 2002 is found in the table below. These numbers can also be found in the instructions accompanying your tax form. They increase every year as well.

Standard Deduction	
married filing jointly	$7,850
single head of household	$6,900
single	$4,700

There are a number of people out there taking the standard deduction. However, with just a little bit of effort, they could be saving themselves a lot of money by itemizing. The more money you deduct from your taxable income, the lower your tax will be.

A quick way to get a rough idea of whether you should be itemizing your deductions or not is to add up the amount of the following items (the biggest and most common itemized deductions) and see if this amount exceeds your standard deduction:

- state income taxes
- property taxes
- mortgage interest
- charitable gifts

If this number is more than your standard deduction, then you should be itemizing. If it is less or about the same, then take the standard deduction. Remember, though, there are many expenses other than these that can be deducted. This is just meant to give you an idea of when to itemize.

If you have had extraordinary medical expenses or significant business expenses in a particular year, you might be able to deduct some of them from your taxable income. Check the instructions with your tax form or talk to a tax professional to see if you qualify.

PROBLEM

You and your husband have recently bought a new home and are thinking that it might be advantageous to itemize your deductions this year. You want to get a quick idea of whether you should consider itemizing this year. Your mortgage interest was $5,432; your property taxes were $1,020; your state and local taxes were $2,463; and your charitable contributions were $1,850. Should you itemize?

STEP 1

Add your mortgage interest ($5,432), your property taxes ($1,020), your state and local taxes ($2,463), and your charitable contributions ($1,850).

$5,432 plus $1,020 plus $2,463 plus $1,850 = $10,765

STEP 2

Get the appropriate standard deduction from the Standard Deduction table on page 171.

married filing jointly = $7,850

STEP 3

Compare the results of Steps 1 ($10,765) and 2 ($7,850). If the amount of your deductions from Step 1 is *greater* than the standard deduction from Step 2, then itemize. If the amount of your deductions from Step 1 is *less* than the standard deduction from Step 2, then take the standard deduction.

The quick deduction tally from Step 1 of $10,765 is greater than the standard deduction from Step 2 of $7,850.

SOLUTION

It is time to itemize.

(Don't worry. It won't be that bad and will save you some money. Just follow the instructions carefully. And if you run into problems, you can always call in a professional.)

 There are expenses other than the ones in the previous example that can be deducted. If you are uncertain, check with the IRS or a tax professional.

You cannot take both the standard deduction *and* itemize your deductions. You have to choose.

In a Nutshell

TAX BASICS
Using Tax Tables

If your taxable income is *less than $100,000*, look up your taxable income in the tax tables that come with your tax form, and find the appropriate tax.

If your taxable income is *$100,000 or more*, follow these steps:

1. Find your taxable income in the IRS tax rate table that relates to your filing status (married filing jointly, single,

continued

continued from previous page

etc.). Follow that line across and follow the directions for your income level.

For example, if your taxable income is $120,000, and you are married and filing jointly, the instructions will read: "the tax is: $24,265.50 plus 30% of the excess over $112,850."

2. Excess over $112,850 = $120,000 − $112,850 = $7,150

3. Change the percent from #2 into a decimal by dividing by 100. 30% = 30 ÷ 100 = .30

4. Tax = $24,265.50 + (the result of #2 × the result of #3) = $24,265.50 + (.30 × $7,150) = $24,265.50 + $2,145 = $26,410.50

 Make sure you do the steps in the parentheses first.

Should You Itemize Your Deductions?

To get a rough idea of whether you should itemize your deductions:

1. Add up the amount of the following items:
   - state income taxes
   - property taxes
   - mortgage interest
   - charitable gifts

2. If the result of #1 exceeds the standard deduction below, itemize.

3. If the result of #1 is lower than the standard deduction below, take the standard deduction.

   Standard Deduction

married filing jointly	$7,850
single head of household	$6,900
single	$4,700

continued

continued from previous page

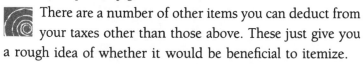 There are a number of other items you can deduct from your taxes other than those above. These just give you a rough idea of whether it would be beneficial to itemize.

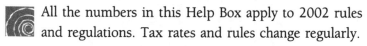 All the numbers in this Help Box apply to 2002 rules and regulations. Tax rates and rules change regularly.

Let's face it. Tax is not the most scintillating subject. If you start talking about tax formulas at a cocktail party, you might well find yourself standing alone. However, taxes do have a profound effect on our disposable income as well as our long-term investment potential. (Considering taxes in investment decisions will be covered in chapter 10.) The more we have a handle on taxes, the better off we will be.

9 Balancing Your Checkbook

The question of why you should keep an accurate, up-to-date checking account comes up all the time. If you want to know your balance, so the argument goes, you can just call the bank or check at the ATM. The bank keeps track of all this stuff. Why should you?

First of all, the bank can only give you your account balance for checks that have already cleared. (That means the checks have come back to the bank, and money has been sent to whomever you wrote the checks to.) The bank has no knowledge of checks you might have written that will clear later that afternoon or the next day or the next week. Suppose your ATM says you have $1,250 in your checking account, and you write a check for that new living room set for $950, thinking you have $300 ($1,250 minus $950) in your account to spare. Then, the next day, three checks you wrote last week totaling $425 clear the bank. Guess what? Some checks are going to start bouncing.

The other thing to keep in mind is that banks make mistakes, and it might happen with your money. In addition, you might find out someone has been using your account without your permission. When balancing my checking account a couple of years ago, I discovered that someone other than I, wrote $1,250 worth of checks. After some investigation, I found that someone had stolen some checks out of the back of my checkbook and bought some very cool stuff with *my* money.

Knowing how much money you have available to spend is simply a really good idea and a very good first step to sound financial management.

Convinced? Then just follow the instructions below. We will first walk through how to keep an up-to-date checkbook registry and then tackle how to reconcile your bank statement.

Keeping a Checkbook Registry

The trick to keeping a checkbook registry is keeping it up-to-date as you go. If you get behind in recording items, it's not the end of the world, but it does make things more difficult. Let's see how it works.

PROBLEM

Your New Year's resolution is to get your financial house in order. You decide the place to start is by keeping an up-to-date, accurate checking account. You have started a new checking account with $1,500. So far, so good. Now what do you do?

STEP 1

Consider yourself forgiven for all past checking account transgressions. (If you need professional religious intervention for this step, get it.)

STEP 2

Record your beginning balance in the balance column of the check registry. (See Table 9-1.) If this is a new checking account, just record how much money you started the account with. If it's an existing account, start with the bank's balance, but be careful, because it is most likely too high. Keep checking with the bank for a while and look at your next statement to make sure you have reflected all your old checks in your balance.

$1,500.00

If you are keeping a check registry for the first time, it might be easier just to open a new account and get a fresh start. Make sure you keep enough money in your old account, however, to cover all the checks you have written and the withdrawals you have made.

STEP 3

Record all transactions made to your checking account, from this day forward, in the appropriate column of your checking registry. Include a description of the transaction in the shaded area. Include checks, ATM withdrawals and any associated fees, telephone transfers, debit card transactions and any associated fees, deposits, and any other money coming into or out of your checking account for any reason.

Don't forget those transactions that are made directly into and out of your account, like direct-deposited paychecks and automatic payment of certain bills.

STEP 4

After you record an individual transaction in your check registry, either add the transaction amount to, or subtract the transaction amount from, your previous balance. Record this new balance in the "Balance" column. You should *subtract* from your previous balance any amount of money going *out* of your account, like checks you write, fees, automatic withdrawals, ATM withdrawals, or debit card transactions. You should *add* to your previous balance any money going *into* your account, like deposits.

Table 9-1: Check Register

Number	Date	Transaction	Payment/Debit–	Fee	Deposit/Credit+	Balance
						$1,500.00
100	1/03/03	Greater Metropolis Gas gas bill	$82.55			$1,417.45
101	1/04/03	Kitty's Cat Care vet	$64.32			$1,353.13
	1/08/03	Be Safe Insurance automatic w/drawal—ins	$30.00			$1,323.13
	1/10/03	ATM cash	$250.00			$1,073.13
102	1/10/03	Phil's Pharmacy misc toiletries	$23.67			$1,049.46
103	1/11/03	Sammy's Sporting Goods tennis racquet	$45.05			$1,004.41
	1/11/03	Lean Machine Fitness fitness club—debit	$35.00	$2.50		$966.91
	1/15/03	Salary direct deposit—payroll			$820.00	$1,786.91

continued

Table 9-1: Check Register (Continued)

Number	Date	Transaction	Payment/Debit–	Fee	Deposit/Credit+	Balance
	1/17/03	Deposit walk-in deposit at bank			$1,000.00	$2,786.91
104	1/17/03	Stuff-mart school supplies	$37.00			$2,749.91
105	1/18/03	Big Bucks National Bank car payment	$233.00			$2,516.91
106	1/19/03	City Food Bank charitable contribution	$100.00			$2,416.91
107	1/20/03	Big Bucks National Bank credit card payment	$344.00			$2,072.91
	1/29/03	Deposit mail-in deposit			$500.00	$2,572.91

Avoid ATM fees by banking at your own bank. Most banks still will not charge you if you have an account with them.

SOLUTION

You now have an accurate, up-to-date check registry.

Doesn't that feel great?

There are a number of very good computer programs that will keep your checking account for you. They will automatically add or subtract transactions and help you reconcile the account at the end of the month. However, you still have to input all the transactions.

Reconciling Your Bank Statement

Now that you have kept such good records, reconciling your bank statement won't be so bad. I promise. Some banks make it very easy for you by providing a handy worksheet with your monthly bank statement. If you bank at a place that does not, just take out your bank statement and carefully follow the steps in the next example.

PROBLEM

Your New Year's resolution has paid off. You have a month's worth of transactions neatly recorded. You have kept your balance up-to-date. Your ending balance in your check registry shows you have $2,572.91 in your account. Your bank statement arrives in the mail, and you are ready to reconcile your account. It says that you have $2,214.31. Yikes. How do you reconcile those two amounts? Where do you start?

▮ STEP 1

Most bank statements include a section for deposits/credits, one for checks, and one for withdrawals/debits. Start with **Deposits/Credits.**

Statement		
Deposits/Credits		
1/15/03	$825.00	electronic transfer
1/17/03	$1,000.00	DEPOSIT
2/08/03	$3.23	interest

Place a check mark in your check registry by the deposits that the bank statement has included.

▮ STEP 2

Record the amount of interest you received this month on your checking account ($3.23) in your check registry in the Deposit/Credit column. Add the amount of the interest to the ending balance in your check registry ($2572.91).

checking registry balance = $2,572.91 plus $3.23 = $2,576.14

▮ STEP 3

Note and resolve any discrepancies in the Deposits/Credits part of the statement with your check registry. If there are discrepancies you cannot justify, don't be shy about calling the bank. They make mistakes, too.

▮ STEP 4

If there were discrepancies in the Deposits/Credits part of the statement from Step 3, make appropriate changes in your registry and adjust your balance accordingly.

For example, you have recorded a deposit for $820.00, instead of the $825.00 in your bank statement. When you check your deposit slip, you realize the bank is correct. First, you change your registry entry for that deposit from $820.00 to $825.00. But your account balance is still $5.00 less than it should be, since you recorded $5.00 less money than was going into your account. So, to make your balance accurate, you need to add $5.00 to it. You could go through and add $5.00 to every balance from the deposit down, but that's pretty tedious. To make it simple, just add $5.00 to the ending balance in your registry from Step 2 ($2,576.14) and be done with it. You could even make an error note in your registry if you wanted to. If you can't figure out whether to add or subtract a mistake, use the handy table below.

checking registry balance = $2,576.14 plus $5.00 = $2,581.14

Reconciling Table	
If your checking registry . . .	**then . . .**
records a deposit too small	add the difference to your registry ending balance.
records a deposit too big	subtract the difference from your ending balance.
records a withdrawal (check, debit, etc.) too small	subtract the difference from your ending balance.
records a withdrawal too big	add the difference to your ending balance.
does not include interest	add the monthly interest to your ending balance.
does not include fees	subtract the fees from your ending balance.

STEP 5

Move to the **Checks** section of the statement.

Checks

100	$102.55	1/14/03
101	$64.32	1/14/03
103	$45.05	1/15/03
105	$233.00	1/23/03
107	$344.00	1/25/03

Place a check mark in your check registry by all the checks that the statement has included. These have cleared the bank.

There will most likely be a number of checks that have not yet cleared the bank. (Note that the date included next to the check on the statement is the date that the check cleared the account, not the date it was written.)

STEP 6

Note any discrepancies between the Checks part of the statement and your check registry.

STEP 7

If there were discrepancies in the Checks part of the statement, make appropriate changes in your registry and adjust the ending balance in your check registry accordingly.

For example, your records show that check #100 was written for $82.55 to the gas company, instead of the $102.55 shown on the bank statement. You go back and look at your gas bill, and see that the bank was correct that your check was for $102.55. You go back and correct the check registry, but your account balance will still be $20.00 more than it should be ($102.55 minus $82.55). To account for the difference, you need to *subtract* $20.00 from your ending balance from Step 4 ($2,581.14) to fix your error. (Use the handy Reconciling Table above, if you can't figure out whether to add or subtract from your balance.)

checking registry balance = $2,581.14 minus $20.00 = $2,561.14

■ STEP 8

Move to the **Withdrawals/Debits** section of the statement.

Withdrawals/Debits			
1/08/03	$30.00	automatic transfer	Be Safe Insurance
1/10/03	$250.00	ATM	
1/11/03	$37.50	point-of-sale debit	Lean Machine Fitness
2/08/03	$7.50	monthly fee	

Place a check mark in your check registry by the withdrawals and debits that are included on the statement. Note that the debit-card transactions reflect the amounts of the debit and the fee added together.

■ STEP 9

Note any discrepancies between the Withdrawals/Debits part of the statement and your check registry. Make appropriate changes to your check registry, and adjust the ending balance on your check registry accordingly. Be sure to include fees and automatic withdrawals you might not have recorded. In this case, you would subtract $7.50 for fees from your ending balance from Step 7 ($2,561.14).

$2,561.14 minus $7.50 = $2,553.64

■ STEP 10

Using your check registry, make a list of the amounts of any transactions you have made but that have not yet shown up on your statement. (These are the transactions that don't have check marks by them.) Make a "−" column and a "+" column. Put the checks and other withdrawals from your account in the "−" column. Put deposits and other additions in the "+" column. Add them up.

	−		+
Check 102	$23.67	Mail deposit	$500.00
Check 104	$37.00	TOTAL	$500.00
Check 106	$100.00		
TOTAL	$160.67		

■ STEP 11

Now take a look at the ending balance in your check registry from Step 9 ($2,553.64) and the ending balance on your bank statement ($2,214.31).

ending balance—checking registry = $2,553.64

ending balance—bank statement = $2,214.31

Don't worry if they are different. They should be.

■ STEP 12

From the ending balance on your bank statement ($2,214.31), subtract the total from the "−" column ($160.67) from Step 10 and add the total from the "+" column from Step 10 ($500).

$2,214.31 minus $160.67 plus $500.00 = $2,553.64

This step evens out the two balances by subtracting the checks that have not cleared and adding the deposits that have not cleared.

 Remember, your *real* balance is the one in your checking account registry. That accounts for all the transactions you have made, regardless of whether they have cleared the bank or not.

Everyone has his own special way of keeping his checking registry and reconciling his checkbook. This is only one of

many ways to approach it. If you can cover the same steps in
a way that makes more sense to you, go for it.

SOLUTION

**The result of Step 12 ($2,553.64) should equal the
amount of your ending balance from Step 9 ($2,553.64)
from your check registry.**

Pat yourself on the back, and congratulate yourself on a job
well done.

If you can't get the two balances to reconcile, don't despair.
There are two approaches. First, if the discrepancy is relatively
small, say a few dollars, you might not feel that it's worth your
time to dig around to try to find the cause. If, after adjusting for
items that have not cleared, your balance on your statement is
larger than your balance on your checking registry, just add the
difference to your registry, and forget about it. If your balance on
your statement is smaller, then subtract the difference from your
registry balance, and forget about it.

If the discrepancy is large, or if it's just bugging you that things
are not coming out, it's time to do some detective work. Carefully
recheck all the checks, deposits, and withdrawals. Pay special attention to fees and automatic withdrawals that you might have forgotten to include. Call the bank if you have to. If you find a mistake,
reflect it in your ending check registry balance. Use the Reconciling
Table on page 184 to help you out.

In a Nutshell

CHECKBOOK
Keeping a Checkbook Registry

1. Record your beginning balance. If starting a new account,
   use the money you started the account with. If you are

continued

continued from previous page

working with an existing account, start with the bank's balance for your account, but know that it is an artificial number (probably too high) because some checks and deposits might not have cleared yet. As old transactions that you have not recorded come to your attention, keep adjusting the ending balance until the old transactions have all cleared the bank.

2. Record all transactions made to your checking account in the appropriate column of your registry. Include checks, ATM withdrawals, automatic withdrawals, telephone transfers, debit-card transactions, fees, deposits, and any other money going into or out of your checking account.

3. For each transaction, add the transaction amount to, or subtract the transaction amount from, your previous balance. Record this new balance in the "Balance" column.

Reconciling Your Bank Statement

1. Carefully check each section of the statement, including "Deposits," "Checks," and "Withdrawals." Place a check mark on your check registry by each transaction that was listed on your statement. That means that transaction has "cleared" at the bank. There will most likely be a number of transactions that you have recorded on your check registry that the bank does not yet have a record of.

2. Make sure the amount of each transaction in the statement exactly matches the amount of each transaction you recorded in your check registry. Remember to record any additional interest or fees that you might not have recorded in your registry, and adjust your ending balance accordingly. Resolve any differences with the bank.

continued

continued from previous page

3. After you have resolved any differences between the transactions in your registry and the bank statement, make appropriate changes to your check registry. If there is a discrepancy in a transaction between your check registry and the bank statement, adjust the transaction in your check registry, and either add or subtract the amount of the difference from the ending balance. If you need help making those adjustments, use the handy Reconciling Table on page 184.

4. Using your check registry, make a list of all those transactions that do not have check marks by them (the ones that have not yet cleared the bank). Make a "+" column and a "−" column. Put checks and other withdrawals in the "−" column. Put deposits and other additions in the "+" column. Add each column.

5. From the ending balance of your bank statement, subtract the total from the "−" column from #4 and to it add the "+" column total.

6. The result of #5 should equal the ending balance from your check registry. If so, reconciliation ends here.

7. If the two balances don't reconcile, do either of the following:
   • If the discrepancy is small, and you don't care about finding it, just add or subtract it from the ending balance of your check registry, and forget about it. (If the result of #5 is larger than your check registry balance, add the difference to your registry balance. If it is smaller, subtract the difference.)
   OR:
   • Carefully recheck all the transactions and all your calculations. Pay special attention to fees and automatic withdrawals that you might have forgotten to include.

While keeping an up-to-date checking account is tedious, keeping track of your money is an essential ingredient in sound financial management. Now that you know how much money you have, you can move on to the next chapter, which will explain how to invest it wisely.

10 Investment Basics

The world of finance has gotten very complicated, with more options out there for individuals than ever before. We are all supposed to be able to make complicated decisions about our "portfolios," when most of us struggle just to balance our checkbook at the end of the month. And then there's the mysterious language used, including words like *T-bills, no-load mutual funds, ARMs,* and *CDs.* I'm still trying to get used to not saying "album." Oops, wrong kind of CD. Anyway, you get the point.

Actually, financial decisions are really not so difficult if you know the basic building blocks to managing your personal finances. This chapter will first explain what interest is, how it is calculated, the difference between simple and compound interest, and how money grows over time with compound interest. We will then go over how to calculate your return (that's the money you earn) on some basic kinds of investments, including onetime contributions and periodic contributions like IRAs, 401(k)s, etc. The chapter concludes by giving you a primer on how to consider taxes in your investment decisions.

Interest

Interest is one of the most important concepts in an understanding of how money grows over time and how to make wise investments.

Simple Interest

While it sounds complicated, interest is actually pretty simple (pun intended). Interest is just an amount of money someone pays another person for use of her money for a while. If you are the one using someone else's money, then you are the one *paying* interest. If someone is using your money, then you are *receiving* interest.

Let's say that you lend cousin Sue $100. You tell her that she can use your money for a year if she returns the $100 to you at the end of the year with $10 interest. At the end of the year, you hope, you'll get $110 ($100 plus $10).

A long time ago, someone decided that the easiest way to compute interest was to charge a certain percentage for the use of someone's money. Let's go back to cousin Sue. When you decided to charge her $10 for use of your $100, you actually charged her 10 percent interest:

($10 divided by $100) times 100 = .10 times 100 = 10%

 A percent is just a number divided by 100.

The same thing happens when you invest money in a bank. The bank gives you some interest for the right to use your money for a while. Generally, interest rates are stated on an annual basis. To calculate how much simple interest you will earn in a year, you can use the following equation:

simple interest = the amount of money invested times percentage of interest

Often the amount invested is called the *principal*, and the percentage of interest is called the *annual interest rate*. So, you might see the formula written this way:

simple interest = principal times the annual interest rate

Let's see how this works by taking a simple example.

PROBLEM

Your son is buying his first car and has saved up all but $2,000. He asks you for a loan. Trying to get him used to the ways of the real world, you agree to loan him the $2,000 for one year at 5% interest. How much should he pay you back at the end of the year?

STEP 1

Change the annual interest rate you are charging him (5%) into a decimal for ease in handling.

5% = 5 divided by 100 = .05

STEP 2

To determine how much interest your son will pay, multiply the principal ($2,000) by the interest rate in decimal form from Step 1 (.05).

$2,000 times .05 = $100

STEP 3

Add the principal ($2,000) and the interest ($100) to determine how much your son will owe you at the end of the year.

$2,000 plus $100 = $2,100

SOLUTION

Your son will owe you $2,100 at the end of the year.

Quick Calculation

A faster way to calculate the total amount of money you earn with simple interest is to use the following:

total money earned = (1 plus annual interest rate) times principal

interest rate = % interest charged

Let's look again at your son's $2,000 loan at 5% interest, and let's do it the quick way this time.

■ STEP 1

To get the interest rate in decimal form, divide the % interest (5%) by 100.

5% = 5 divided by 100 = .05

■ STEP 2

Add 1 to the result of Step 1 (.05).

1 plus .05 = 1.05

■ STEP 3

Multiply the result of Step 2 (1.05) by the principal.

1.05 times $2,000 = $2,100

■ SOLUTION

Your son will still owe you $2,100 at the end of the year.

Compound Interest—How Money Grows over Time

In theory, compound interest is no different than simple interest, except that the interest and principal are being calculated on an ongoing basis rather than just at the end of the year. By calculating the interest and adding it to the principal more often, the financial institution that holds your money gives you a much better return on your investment. Let's see how this works.

You put $10,000 in a 10-percent-interest savings account. At the end of the year, the bank gives you the entire 10 percent interest, or $1,000 (10% of $10,000 = (10 ÷ 100) times $10,000 = .01 times $10,000 = $1,000).

Let's say another bank also offers you 10 percent interest, but it gives it to you in two parts, half in the middle of the year and the other half at the end of the year. So, you would receive 5 percent (½ of the 10% interest) after six months and the other 5 percent at the end of the year. The table below shows you how your $10,000 would grow.

First, change 5 percent interest into .05 by dividing 5 by 100.

at the end of	old balance		interest rate	interest	new balance (old balance plus interest)
6 months	$10,000	times	.05 =	$500	$10,500.00
1 year	$10,500	times	.05 =	$525	$11,025.00

You can see by this example that the new balance calculated at the end of the first six months becomes the basis for the second interest calculation at the end of year. *Compounding*, or recalculating principal and interest, two times a year gives you $1,025 in interest ($500 plus $525). That's $25 more than the $1,000 the other bank was going to give you by computing your 10 percent interest just once at the end of the year.

While $25 might not seem like much, you can imagine what might happen with larger sums of money over many years, compounded at even smaller intervals—quarterly, monthly, weekly, or even daily. Understanding how money grows over time through compounding is essential to making sound financial decisions. Next we will look at how to put this knowledge to work by looking at some common investment scenarios.

When comparing banks or other sources of interest-bearing accounts, if the services and interest rates are the same, look for the one that compounds most frequently.

In a Nutshell

INTEREST

Simple Interest

simple interest = principal × the annual interest rate

principal = amount of money invested

interest rate = % annual interest (usually converted to a decimal by dividing % value by 100)

final amount of money = principal × (1 + annual interest rate)

Compound Interest

compound interest = the amount of interest you acquire by recalculating principal and interest at regular intervals

Investments

Using a Financial Calculator

There is a complicated formula that calculates how money grows over time and how much money you can earn over a variety of time frames and with varying interest assumptions. But why use a complicated formula when there are lots of tools out there to do the work for you? Financial calculators that will do this calculation, as well as a number of other interesting calculations, are pretty cheap these days. There are even reliable, easy-to-use Internet sites that have financial calculators that allow you to make all these calculations for free. And some of the new, handheld Palm Pilots have these calculations built right in. (I did all the calculations for this chapter and the next with a $30 calculator, then checked them all on a free Internet site.)

Unfortunately, each financial calculator has its own terminology and way of doing things. Do some deep-breathing exercises, then take a minute to look at the user's manual. It usually runs through how to do basic calculations. The sections that follow will give general advice and some tips on how to use a financial calculator or similar tool to calculate how much your money will be worth in some common investment scenarios.

To invest wisely, it is very important that you know the complete terms of a prospective investment, including the time frame of the investment, the minimum investment necessary, the expected return and whether that return is guaranteed, and any conditions or limitations put on the investment.

How Much Will Your Money Be Worth in the Future?

Onetime Contribution

Let's start looking at investments with a onetime contribution. Typically, you'll be in the situation where you want to invest a lump sum of money and determine how much it will be worth over a certain amount of time. You might receive an inheritance, sell some stock, get a settlement of some kind, or receive a bonus. To calculate how much your money will be worth in the future with a onetime contribution, you will need to input the following information into a financial calculator or similar tool:

- yearly interest rate you are being offered
- the amount of the principal you are investing
- how long you are going to invest the money

Let's look at the example below to see how this calculation works.

PROBLEM

Your dear old great-uncle Ed died recently and left you $15,000. You decide to put the money in a money market fund to save for the kids' education. The interest rate on the money market is 4.5%. How much will the $15,000 be worth in 12 years, when the first one goes off to college?

STEP 1

List the information you'll need to input into an financial calculator or spreadsheet program.

yearly interest rate = 4.5%

amount of principal = $15,000

how long you are going to invest the money = 12 years

STEP 2

Input the information into the financial calculator or other financial tool.

total money earned = $25,438

SOLUTION

Uncle Ed's $15,000 will be worth $25,438 at the end of 12 years.

(Thanks, Uncle Ed.)

Understanding how money grows with interest over time is the basic building block to making smart financial decisions. In the example above, armed with the information you now know, you could easily compare different investment options for Uncle Ed's $15,000 to see how much each would yield.

If you are investing a certain amount of money over a long period of time, remember, that money might not be worth as much in the future because of inflation. For example, the $100,000 you wanted your investments to yield for your kids' college education probably will not buy as much education in the future as it would today.

Regular-Period Contributions (IRAs, 401(k)s, Savings Plans)
Now that you understand how to calculate the earnings (or yield) on a single contribution, let's move on to investments that are based

on your making a regular contribution each week, month, or year. As in the case of investing a single amount of money, there are plenty of tools out there to help you calculate how much your money will be worth at the end of some period of time if you make regular payments.

You just need the following information:

- yearly interest rate you are being offered
- how long you are going to invest the money
- the amount of the regular contribution
- the number of times per year the contribution is made

Look at the following example to see how much your regular contributions will be worth after a specific amount of time.

PROBLEM

Your daughter has begun a lawn-service operation to earn extra money. You are encouraging her to invest $2,000 of the earnings each year in an IRA. She is seventeen and has other things in mind for the money. You would like to convince her of the rightness of this investment using some cold, hard facts. If she puts $2,000 a year in her IRA for the next 50 years, how much will it be worth it be worth at the end of 50 years? Her IRA will earn 7.5% a year.

STEP 1

List the information you need to input into an financial calculator or other financial tool.

yearly interest rate = 7.5%

how long you are going to invest the money = 50 years

the amount of the regular payment = $2,000

the number of times per year the payment is made = 1

■ STEP 2

Input the information into the financial calculator.

total money earned = $965,000

■ SOLUTION

With an investment of $2,000 a year, your daughter's IRA will be worth $965,000 50 years from now.

(Wow. That should impress her.)

Each financial tool has its own way of doing things. Some require the length of investment in months instead of years, in which case you multiply the number of years of the investment by 12 months in a year. Some require monthly interest rates be inputted instead of yearly interest rates. In this case, simply divide the yearly rate by 12. In all cases, just follow the instructions carefully, and you should not have any problem.

Remember that you will have to pay taxes on the interest earned on most investments. The amount and timing of taxes on investments vary significantly depending on the type of investment. See the section below, "Considering Taxes in Your Investments."

Annual Effective Yield

You will often see interest rates on investments advertised in two ways—annual interest rate and *annual effective yield* (or *annual percentage yield*). The annual effective yield is just the annual interest rate you would receive after taking compounding into effect. For example, a bank which compounds monthly might offer 6.25 percent interest, 6.43 percent annual effective yield. That means if you invest $100 at 6.25 percent compounded monthly, at the end of one year you will actually get $6.43 interest. As we saw in the earlier section on compounding, you get more than the stated interest if you calculate the principal and interest more often than once a year.

If the annual effective yield rate is available, you should use it in your calculations on investment income in place of annual interest rate, since it is a more precise indicator of how much money you will actually make. You will get to exactly the same point, however, if you have a calculator that allows you to specify how often your investment will compound. In that case, you can just input the interest rate as well as the number of times a year the interest will compound.

If you are using annual effective yield as your interest rate, do not specify anything other than a yearly compounding, or you will, in effect, be double-counting.

In a Nutshell

INVESTMENTS
How Much Will Your Money Be Worth in the Future?

Onetime Contribution

1. Use a financial calculator or other financial tool to calculate how much a specific amount of money will be worth at the end of a certain period of time.

2. Input the following:
   - yearly interest rate
   - amount of principal
   - how many years you are going to invest the money

3. Using your inputs, the financial calculator will calculate how much your investment will be worth.

continued

continued from previous page

Regular Periodic Contributions (401(k)s, IRAs, Savings Plans)

1. Use a financial calculator or other financial tool to calculate how much money you will earn over a certain amount of time if you are making regular contributions.

2. Input the following:

    • yearly interest rate

    • amount of regular payment

    • how many times a year the contribution is made

    • how many years you are going to invest the money

3. Using your inputs, the financial calculator will calculate how much money you will earn with your regular contributions over a certain time.

Considering Taxes in Your Investments

Considering investments on an after-tax basis certainly adds a layer of complexity to your decision making. But the truth of the matter is that we all pay taxes, and you will get a much more accurate understanding of different investment options if you consider their tax implications. This section will walk you through how to calculate your after-tax return on several common types of investments. With this knowledge in hand, you can then more effectively evaluate different investment options.

You need to know two things before you start looking at investments on an after-tax basis: how to calculate both your marginal tax rate and your capital gains tax rate.

All the tax calculations in this chapter are based on 2002 rules and regulations. Up-to-date tables, formulas, deductions, and other tax information are in the instructions that accompany your tax form.

Marginal Tax Rate

The first step to evaluating investments on an after-tax basis is to know your *marginal tax rate*. Your *marginal tax rate* is the tax rate that would apply to the very next taxable dollar you would earn. To find your marginal tax rate, just take your gross income, add one dollar to it, and look in the table below to see the rate at which it is taxed.

Estimating Your Marginal Tax Rate

Find the column that has your filing status. In that column, find the lowest number that is still higher than your gross income. Follow the row with that number over to the far right to find your estimated marginal tax rate.

Married Filing Jointly	Single Head of Household	Single	Your Estimated Marginal Rate
$ 25,850	$ 22,900	$ 13,700	10%
$ 60,550	$ 50,350	$ 35,650	15%
$126,700	$109,600	$ 75,400	27%
$185,800	$169,500	$148,950	30%
$320,900	$319,950	$314,750	35%
+	+	+	38.6%

For example, if you are married and the two of you make $150,000 gross income, you would choose $185,800, since it is the lowest number in the Married Filing Jointly column that still exceeds your $150,000 gross income. Follow the row with $185,800 over to the far right to find that your estimated marginal tax rate is 30 percent.

Capital Gains Tax Rate

The second tax rate that you need to be familiar with is the *capital gains tax rate*, which is a different tax rate for certain types of investments (We will go over which ones capital gains tax applies to in the next section). Your *capital gains tax rate* is based on your marginal tax rate. You can find your capital gains tax rate in the table below.

Estimating Your Capital Gains Tax Rate	
Marginal Rate	**Capital Gains Rate**
less than or equal to 15%	10%
greater than 15%	20%

For example, if your marginal tax rate is 30 percent, your capital gains tax rate would be 20 percent. If your marginal tax rate is 15 percent, your capital gains tax rate would be 10 percent.

Taxes on Different Types of Investments

Now that you know your marginal tax rate and your capital gains tax rate, take a look at the table below to see how some common investments are taxed.

Taxes on Investments
• Investments taxed as regular income—interest on CDs, interest on money market accounts or savings accounts, stock dividends, corporate bonds, and the like are taxed at the same rate as your salary. Use your *marginal tax rate* to determine the tax on these investments.
• Investments taxed as capital gains—profits you make by owning property for more than a year and selling it; this includes real estate property as well as stocks. Use your *capital gains tax rate* to determine the tax on these investments. If you hold the property for less than a year, the gain is taxed as ordinary income, and you should use your marginal tax rate instead. (There is generally an exemption from capital gains on selling your home; talk to your tax professional about this.)
• Investments that are tax-exempt—interest you make on most state and local government bonds are not subject to taxes.
• Investments that are tax-deferred—profits and interest you make on most retirement plans, including 401(k)s, IRAs, etc., are not taxed until you withdraw from the plans; the assumption is that your tax rate will be much smaller after retiring. (Penalty taxes can apply to withdrawals before age fifty-nine and a half.)

Computing After-Tax Rate of Return on Investments

Now that you know how to estimate both your marginal tax rate and your capital gains tax rate, you are ready to estimate your *after-tax rate of return*, which is the percentage interest or percentage yield you will receive on an investment after you pay taxes. Understanding the tax implications of investments will give you a much better idea of how they will actually perform. To estimate your *after-tax rate of return*, follow the steps below:

1. Determine the kind of investment you have and whether it is taxed at the capital gains rate or the marginal tax rate or not taxed at all, using the Taxes on Investments table above.
2. Look at the Estimating Your Marginal Tax Rate table on page 205 to determine your marginal tax rate, and maybe the Estimating Your Capital Gains Tax Rate table (page 206) to determine the capital gains rate, depending on the type of investment.
3. Change the tax rate (either capital gains or marginal) into a decimal by dividing it by 100.
4. Calculate the investment multiplier:
   multiplier = 1 minus (marginal tax rate or capital gains tax rate)
5. Calculate the after-tax rate of return:
   after-tax rate of return = (pretax interest rate, or rate of return) times multiplier

Let's walk through an example.

PROBLEM

You have a CD that promises 5% interest, and you are interested in determining its after-tax rate of return. Your gross income is about $65,000. You are single. What is your after-tax interest rate, or rate of return?

STEP 1

Look at the Taxes on Investments table on page 206 to determine how your CD is taxed.

Your CD is taxed as regular income, so you use your marginal tax rate.

STEP 2

Find your estimated marginal tax rate on the Estimating Your Marginal Tax Rate table on page 205.

Your $65,000 salary will have an estimated marginal tax rate of 27%.

STEP 3

Convert your estimated marginal tax rate from Step 2 (27%) into a decimal by dividing it by 100.

27% = 27 divided by 100 = .27

STEP 4

Calculate your investment multiplier by subtracting from 1 the estimated marginal tax rate in decimal form from Step 3 (.27).

investment multiplier = 1 minus .27 = .73

STEP 5

Calculate your after-tax rate of return by multiplying your expected return (5%) by the investment multiplier from Step 4 (.73).

after-tax return rate = 5% times .73 = 3.65%

SOLUTION

The after-tax rate of return on your CD is 3.65%.

Now that you know how to calculate the after-tax return on different kinds of investments, you can now get a good estimate of how much a specific investment will be worth in the future. Use the techniques described earlier in this chapter, in the section "How Much Will Your Money Be Worth in the Future?" Just substitute an after-tax rate of return for the interest rate on your investment. For example, if you wanted to know how much your CD in the above example would actually return after ten years, taking taxes into account, you would input "3.65%" into your financial calculator as the interest rate.

Comparing Investments on an After-Tax Basis

You now have all the tools you need to compare investments on an after-tax basis. Just calculate the after-tax rate of return on each, and all other things being equal, choose the one with the highest rate of return.

PROBLEM

You are considering two different investment options. The first is a tax-exempt municipal bond that will yield 5% interest. The second is some stock that will be subject to capital gains tax if you sell it. You expect a return of about 6% on the stock (although you know that past performance is no guarantee of future performance). You and your wife together make $85,000. What is your after-tax rate of return on these two investments, and which is the better investment?

STEP 1

Starting with the first investment, compute your after-tax rate of return. Since the first investment is tax-exempt, the after-tax return rate is the same as the pre-tax return rate (5%).

tax-exempt municipal bond after-tax return rate = 5%

STEP 2

Look at the Taxes on Investments table on page 206 to determine how stocks are taxed.

Profits from selling stocks are taxed at capital gains rates.

STEP 3

To determine your capital gains rate, you first have to calculate your marginal tax rate. Look at the Estimating Your Marginal Tax Rate table on page 205 to estimate your marginal tax rate.

Profits from selling stock and selling stock dividends are taxed differently. Tax dividends are taxed at regular rates. Profits from selling stock held over a year are taxed at capital gains rates.

Since you are married, and you make less than $126,700, your estimated marginal tax rate is 27%.

STEP 4

Determine your capital gains rate based on your marginal rate from Step 3 above (27%). Use the Estimating Your Capital Gains Tax Rate table on page 206.

Since your marginal tax rate is above 15%, your capital gains tax rate is 20%.

STEP 5

Change your capital gains tax rate of 20% into a decimal by dividing it by 100.

20% = 20 divided by 100 = .20

STEP 6

Calculate the investment multiplier by subtracting from 1 your capital gains tax rate in decimal form from Step 5 (.20).

investment multiplier = 1 − .20 = .80

STEP 7

To calculate the after-tax return on the stock option, multiply your expected return rate (6%) by your investment multiplier from Step 6 (.80).

after-tax return on the stock option = 6% times .80 = 4.8%

STEP 8

Compare the returns on the two investments you are considering, from Steps 1 and 7.

tax-exempt municipal bond return = 5%

stock after-tax return = 4.8%

SOLUTION

All other things being equal, you should pursue the tax-exempt bond that returns 5% after tax.

Notice that the amount of the investment does not matter. You need only compare the after-tax rates of return.

It's important to make investment decisions based on *after-tax* returns. As the above example demonstrates, the most attractive pretax rate of return does not always amount to the biggest return on your investment after you pay your taxes.

The process above is meant to be just an *estimate* of your after-tax rate of return to give you a rough idea of how to compare different investments. To be more precise, you would have to include the length of the investment, the effect of state taxes (which are sometimes quite high), fees, and a number of other variables. At that point, you might want to consult a financial advisor.

In a Nutshell

CONSIDERING TAXES IN YOUR INVESTMENTS
Computing After-Tax Rate of Return on Investments

1. Determine the kind of investment you have and whether it is taxed at the capital gains rate or the marginal tax rate or not taxed at all, using the Taxes on Investments table on page 206.

2. Look at the Estimating Your Marginal Tax Rate table on page 205 to determine your marginal tax rate, and maybe also the Estimating Your Capital Gains Tax Rate table (page 206) to determine the capital gains rate, depending on the type of investment.

3. Change the tax rate (either capital gains or marginal) into a decimal by dividing it by 100.

4. Calculate the investment multiplier:
   multiplier = 1 minus (marginal tax rate or capital gains tax rate)

5. Calculate the after-tax rate of return:
   after-tax rate of return = (pretax interest rate or rate of return) times multiplier

Comparing Investments After-Tax

Compute the after-tax rate of return for each investment, and all other things being equal, choose the investment with the highest return rate.

Now that you know the basics of investing, you can explore the more complicated financial calculations that are beyond the scope of this book. Even if future financial decisions might at times be confusing, know that you are armed with a good handle on the basics. And you can always return to this chapter if you need a review.

11 Borrowing Money

R emember how in chapter 10 we defined interest as an amount of money someone pays another person in order to use her money for a while? In that chapter, we reviewed a number of scenarios in which you were *receiving* interest. With loans, the principles involved are essentially the same, except you are the one *paying* the interest.

With a loan, you normally make an agreement with a bank or some other lending institution to pay back a specific amount of money within a specific time frame, at a specific rate of interest, at a specific level of payment. As was the case with receiving interest, there is a formula that can calculate any of these "specifics" for you. And again, there are financial calculators, Internet sites, and other financial tools available that will do all the work for you. So let them.

In this chapter, we will review the basics of borrowing money, including how to calculate what your monthly payments will be when you take out a loan. This chapter will also explain how to calculate the real interest you are paying on those "low-cost" loans. It is probably more than you think. We will also go over mortgage payments and how to compare fifteen-year and thirty-year mortgages. That topic we all love to hate, credit cards, will also be covered in depth. Finally, this chapter will explain how to consider taxes in your loan decisions.

Calculating Monthly Payments—Car Loans, Furniture, Etc.

When you take out a loan, you most often pay it back on a monthly basis, paying the same amount of money each month until the loan is paid off. A process called *amortization* takes the total amount of money you owe over the life of a loan, including interest, and splits it into equal, monthly payments. The *principal* is the amount of the payment that goes directly to pay off the loan. The interest part of the payment pays the interest on the loan.

To calculate the monthly payment you will need to make to pay off a loan of a certain amount, just use your financial calculator or other financial tool, as you did in chapter 10, "Investment Basics." Virtually every financial calculator will have a function that gives you the monthly payment on a loan after you input the following information:

- amount of loan
- interest rate on the loan
- total number of years you have to pay off the loan

Look at the following example to see how you would use your financial calculator to compute monthly loan payments.

PROBLEM

You have had your eye on this wonderful little sports car for years. You have had a good year financially and decide it is time for you to have it. You find a used model in good condition for $22,500. Your credit union is offering 7.0% interest on used cars with a 5-year loan. You have saved $5,000 that you will put towards the purchase of the car. How much will your monthly payments be?

STEP 1

Determine the amount of the loan by subtracting your contribution ($5,000) from the total price of the car ($22,500).

$22,500 minus $5,000 = $17,500

STEP 2

List the factors you will need to input into your financial calculator or other financial tool.

amount of loan = $17,500

interest rate on loan = 7.0%

number of years of the loan = 5

STEP 3

Input the factors from Step 2 into your financial tool, and ask it compute the amount of the monthly payment.

monthly payment = $347

SOLUTION

Your monthly payment for five years will be $347.

Credit unions, banks, and car dealers offer a variety of different kinds of car loans. Also, interest rates on car loans often vary between new and used cars. Make sure you know all the details of a loan for the type of car you are buying.

Many lending institutions will offer a very attractive beginning interest rate but after a certain amount of time will increase the rate dramatically. Be careful to check out carefully the terms of the loan and shop around to find the loan with the best terms for your situation. Be sure to look at the time period of the loan, the interest rate throughout the entire life of the loan, and how often the interest is calculated (daily, monthly, quarterly, etc.).

Often a financial tool will ask you for the number of payments in a year or the number of payments over the life of the loan. Generally, loan payments are made once a month, so the number of payments in a year would be 12. To calculate how many payments over the life of the loan, just multiply the number of years in the loan by 12. For example, if a loan is for 5 years, you will generally make sixty (12 times 5) payments.

The payment calculated by a financial tool only includes the principal and the interest. If taxes, insurance, or other fees are charged, you need to add them to the payment.

Calculating the Real Interest You Are Paying

In the above section, you calculated what your monthly payments would be on a loan, given the amount of the loan and the annual interest rate. Often, loans on consumer goods, like furniture, do not advertise the annual interest rate outright, but do include an estimate of the "low monthly payments" you would have to make. To see if a financing plan is a good deal, it is best to know what interest you are being charged, based on the monthly payments you would be required to pay. You can use the same financial tool you have been using with other financial calculations.

To calculate interest, you will need to know:

- loan amount
- number of years of the loan
- amount of monthly payment

PROBLEM

You see an ad in the paper for that great dining room set you have been coveting at the Chic Shack. The price is $2,500, but you can finance it over 5 years for $70 a month

with no money down. That doesn't sound too bad. But you're wondering, how much interest is the Chic Shack actually charging you?

STEP 1

List the items you will need to input into your financial calculator or other financial tool.

loan amount = $2,500

number of years of loan = 5

amount of monthly payment = $70

STEP 2

Ask the calculator to compute the annual interest rate, given the input from Step 1.

annual interest rate = 23%

SOLUTION

The Chic Shack is charging you 23% interest.

(Wow. That interest rate is really high. You might look at other ways to finance the furniture. Better yet, just save up until you can pay for it without credit.)

A quick and easy way to determine whether or not you should buy on credit is to determine the real cost of buying the item after you make all the payments. In the above example, you pay $70 a month for 5 years (60 months). That means the dining room set that was priced for $2,500 actually would cost you $4,200 ($70 times 60 months). That is $1,700 more than the sales price ($4,200 minus $2,500). Ouch.

Mortgages

Mortgages work the same as the other loans we have just gone over. They just have a long payoff period and are generally used to finance the purchase of your home. For fixed-rate mortgages (the ones where the interest stays the same for the whole loan) you can calculate payments and interest just the way you would with a car loan or any other loan.

Each month, your mortgage payment most likely pays a number of things, including taxes and various types of insurance. The bulk of your payment, however, goes to the two parts of paying off a loan: *principal* and *interest*. The *principal* is the part of the payment that goes directly to paying off your loan. The *interest* part of the payment pays the interest on the debt. Most mortgage statements will give you a detailed breakdown of your mortgage payment as well as a report of your "principal balance." Your *principal balance* is the amount you still owe on the loan.

It can be pretty scary to be ten years into your mortgage and realize how little of the loan you have paid off. That is because, in a standard home mortgage, interest is front-end loaded. That means that early in the loan cycle, your monthly payment is primarily interest payment, with very little going to pay off the loan itself. Toward the end of the loan cycle, most of the payment is principal and therefore going toward paying off the loan. This effect has important ramifications for taxes, since you can deduct the interest on most mortgages from your taxes, while you cannot deduct the principal. (More on taxes later in the chapter.)

Estimating Your Monthly Mortgage Payment

Your lender will be able to tell you what interest she is offering and what your monthly payments will be, including estimated taxes and insurance payments. If you are considering buying a home and want to get an approximate idea of your monthly payment on your own, use the same technique that we used in the "Calculating Monthly Payments" section of this chapter. Just remember your estimated payment will most likely be a little low because you have not yet added on taxes and insurance.

You just need to input the following information into your financial calculator:

- loan amount
- number of years of the loan (usually 15 or 30)
- interest rate

Let's try an example.

PROBLEM

You and your wife have just found your dream house. You would like to offer the sellers $250,000 for their house. You are prepared to make a down payment of $50,000. Your lender says the bank will give you a 30-year fixed-rate mortgage at 7.5% interest. What will be your approximate monthly payment, excluding taxes and insurance?

STEP 1

Subtract your down payment ($50,000) from the price of the house ($250,000) to determine the approximate loan amount.

$250,000 minus $50,000 = $200,000

STEP 2

List the items you will need to input into your financial tool.

loan amount = $200,000

number of years of the loan = 30 years

interest rate = 7.5%

STEP 3

Input the items from Step 2 into your financial calculator and ask it to give you your monthly payment.

monthly payment = $1,400

◼ SOLUTION

Your monthly payment for a $200,000 30-year fixed-rate loan at 7.5% is approximately $1,400.

Thirty-Year versus Fifteen-Year Mortgage

You can use the same techniques that we have been using to compute monthly payments to compare the monthly payments on thirty-year and fifteen-year mortgages. Often, a home buyer will want to consider a shorter mortgage, so as not to have a large piece of debt for such a long time. Also, the interest rate is often lower with a shorter mortgage.

◼ PROBLEM

In the above example, you are considering paying off your dream house in 15 years instead of the standard 30. Your lender says that the interest rate for a 15-year, fixed-rate mortgage is 7.0%, a full .5% lower than for a 30-year loan. That sounds very attractive. What would your approximate monthly payment be with the 15-year mortgage?

◼ STEP 1

Subtract your deposit ($50,000) from the price of the house ($250,000) to determine the approximate loan amount.

$250,000 minus $50,000 = $200,000

◼ STEP 2

List the items you will need to input into your financial calculator.

loan amount = $200,000

number of years of the loan = 15 years

interest rate = 7.0%

STEP 3

Input the items from Step 2 into your financial calculator and ask it to give you your monthly payment.

monthly payment = $1,800

SOLUTION

Your monthly payment for a 15-year-fixed rate mortgage at 7.0% is $1,800.

(Your monthly payment for a 30-year-fixed rate mortgage at 7.5% is $1,400.)

You can see from the above example that you will pay approximately $324,000 on the fifteen-year loan (15 years times 12 payments in a year times $1,800), and you would pay approximately $504,000 with a conventional thirty-year loan (30 years times 12 payments a year times $1,400). Of course, to that you would have to add taxes and other fees. Not including these, you would save $180,000 ($504,000 minus $324,000) over the life of the loan.

On the face of it, it would seem that the shorter-term mortgage makes better financial sense than a longer-term mortgage. In many cases it does, particularly if a shorter loan serves as a forced savings plan you would not otherwise have. However, given the tax benefits of deducting interest income, sometimes it is best to use the additional funds you would be paying to your mortgage for alternative investments if you have the financial discipline to do so. It is best to talk with your financial advisor about your individual situation. You will get more tips on looking at mortgages—on an after-tax basis—later in this chapter.

Adjustable-Rate Mortgages

A number of financial institutions offer adjustable-rate mortgages, or "ARMs." ARMs do just what the name suggests; they adjust the mortgage interest rate over the life of a loan, usually in relationship to a published index. ARMs usually offer an enticing introductory rate, then revert to adjusting at fixed intervals over the rest of the life of the loan, with a cap that keeps the loan rate from running amok. It is best to talk to your lender about the payments for that kind of a loan.

In a Nutshell

BORROWING BASICS
Calculating Monthly Payments

Use a financial calculator or other financial tool to calculate what your monthly payments would be to pay off a loan, including a mortgage.

1. Input the following:
   - amount of loan
   - interest rate on the loan
   - total number of years to pay off the loan

2. Using your inputs, the financial calculator will calculate your monthly payment.

3. If you are calculating a monthly mortgage payment, this method will most likely underestimate your payment because you have not yet added on taxes and insurance.

Calculating Interest

Use a financial calculator or other financial tool to calculate what interest you are being charged.

1. Input the following:
   - amount of loan
   - total number of years to pay off the loan
   - the monthly payment

2. Using your inputs, the financial calculator will calculate your annual interest rate.

Credit Cards

Credit cards are just another way of borrowing money. The financial institution that holds the credit card lends you the money to make purchases. If you pay the institute back promptly at the end of the billing cycle, it does not charge you any interest on the loan. So far, so good. However, if you do not pay fully every month, it does charge interest, often at a rate well above what you would be charged if you borrowed money in a more conventional way.

Credit cards are generally a convenient and safe way to make purchases. They can also cause serious financial distress if not used wisely. Attractive advertising, special introductory interest rates, and our human inclination to want more stuff all combine to make buying with credit cards an appealing, yet financially dangerous, option.

Credit cards do not have to be such a downer. If you use them carefully and pay off your balance every month, they really do not need to pose any problem for you. In order to help you use them prudently, this section reveals the real cost of buying with a credit card, reviews some of the possible pitfalls of using a credit card, and even helps you get out of trouble if this section has come a little too late for you.

Interest Rates—The Real Story

There is a great variety in credit cards—fees or no fees, varying interest rates, and differing terms and conditions. Given the diversity in choices, you will definitely want to shop around for a card with the most positive terms.

There is one unifying factor, however, among all credit cards. If you do not pay off your balance every month, credit cards are a very expensive way to buy things. The interest rates, or "finance charges," on unpaid balances are often well above those charged for other kinds of loans. Expensive fees charged if you are late with your payment make matters even worse. And with many credit cards, cash advances are subject to finance charges, regardless of whether you paid your balance off the previous month.

Most credit cards these days express interest rates in terms of an annual percentage rate (APR). With lending institutions that calculate interest monthly, you can simply divide the APR by 12 to determine the interest rate on your monthly bill. Be careful, though. I have a credit card that computes interest *daily*.

In chapter 9 we discussed how much more money you could make on your investments if the bank compounded the interest by calculating it more often. Well, credit card companies do the same thing when they collect interest from you. If you keep an unpaid balance on a credit card with an APR of 18 percent that compounds monthly, you are really paying *an effective interest rate* of nearly 20 percent over the course of a year.

If you look at your statement, you will also notice that finance charges are generally based on "average daily balance." Average daily balance is just the amount of money, on average, that you owe the credit card company for the previous month. If you pay off your balance every month, your average daily balance for the next month is 0, and therefore you will not be subject to interest, regardless of the amount of new purchases. However, if you do not pay your bill in full and carry over even a small amount from the previous month, your average daily balance will then, in most cases, include any new purchases you make over the next month's billing cycle. You will then be charged interest on new purchases that you have not even been billed for yet. Therefore, you should make it a practice to pay off your balance fully every month.

Let's say you left $10 unpaid on your balance. On the last day of the thirty-day billing cycle, you charged an item that cost $2,990. Your average balance, on which you would be charged interest, would be ([$2,990 plus $10] divided by 30 days in the cycle =) $100.

Credit cards vary dramatically in terms of their interest rate, calculation of finance charges, fees, etc. It is best to shop around for a credit card that has the best terms.

The Real Price of an Item Bought with a Credit Card

As is the case any time you buy on credit, it is good to know how much you are actually paying for a product. To do that, just add up all the payments you have made on your credit card that pertain to that product.

PROBLEM

Your golf buddies have been giving you a hard time about the old, out-of-date clubs you have been using. (That red mitten you keep over your putter probably doesn't help much.) You've been watching the sales at the Teerific Golf Shop. The clubs you have been dying to have finally went on sale for $1,750, marked down from $1,925. While you can't really afford them, it is such a great sale, you decide you will just put them on your credit card and pay the balance off at about $100 a month. The APR on your credit card is 18%. Your card calculates interest on a monthly basis. How long will it take you to pay the clubs off, and what is the real price of the golf clubs?

STEP 1

(Refer to Table 11-1 below.) Convert APR (18%) to a decimal for ease of use.

18% = 18 divided by 100 = .18

A percentage is just a number divided by 100. To convert a percent to a decimal easily, just place a decimal after the percent and move it two places to the left. If it is only one digit (e.g., 5%), assume a zero in front of the number (5% = 05% = 05.% [moved two decimal places to the left] = .05).

STEP 2

Divide the annual APR in decimal form from Step 1 (.18) by 12 to calculate monthly interest.

.18 divided by 12 = .015

STEP 3

For each month except the first month, you will calculate interest on your credit card balance by multiplying the balance by the interest rate from Step 2. Start with month 2.

month 2: $1,650.00 times .015 = $24.75

(There is no interest on the first month, since that is the first opportunity you have had to pay your bill.)

STEP 4

Add the interest for that month to your balance.

month 2: $1,650.00 plus $24.75 = $1,674.75

STEP 5

Subtract your $100 payment from the balance plus interest from Step Four to calculate your new balance.

month 2: $1,674.75 minus $100.00 = $1,574.75

The new balance from Step 5 becomes your credit card bill for the next month.

STEP 6

To determine the real price of the golf clubs, go through this process for each month, then total all the payments made.

SOLUTION

It took you 21 months to pay off the golf clubs, and you paid $2,009.95 for them.

In reality, you might not spend the time to go through the kind of analysis in the above example. But you can take a few lessons from it nonetheless. In this example, you would have ended up spending approximately $2,010 on clubs that were priced at $1,750.

Table 11-1: Calculating Cost with Interest

Month	Starting Balance	Interest (Finance Charge)	Balance plus interest	Payment	New Balance
1	$1,750.00	no interest	$1,750.00	$ 100.00	$1,650.00
2	$1,650.00	$24.75	$1,674.75	$ 100.00	$1,574.75
3	$1,574.75	$23.62	$1,598.37	$ 100.00	$1,498.37
4	$1,498.37	$22.48	$1,520.85	$ 100.00	$1,420.85
5	$1,420.85	$21.31	$1,442.16	$ 100.00	$1,342.16
6	$1,342.16	$20.13	$1,362.29	$ 100.00	$1,262.29
7	$1,262.29	$18.93	$1,281.23	$ 100.00	$1,181.23
8	$1,181.23	$17.72	$1,198.94	$ 100.00	$1,098.94
9	$1,098.94	$16.48	$1,115.43	$ 100.00	$1,015.43
10	$1,015.43	$15.23	$1,030.66	$ 100.00	$ 930.66
11	$ 930.66	$13.96	$ 944.62	$ 100.00	$ 844.62
12	$ 844.62	$12.67	$ 857.29	$ 100.00	$ 757.29
13	$ 757.29	$11.36	$ 768.65	$ 100.00	$ 668.65
14	$ 668.65	$10.03	$ 678.68	$ 100.00	$ 578.68
15	$ 578.68	$ 8.68	$ 587.36	$ 100.00	$ 487.36
16	$ 487.36	$ 7.31	$ 494.67	$ 100.00	$ 394.67
17	$ 391.67	$ 5.92	$ 400.59	$ 100.00	$ 300.59
18	$ 300.59	$ 4.51	$ 305.10	$ 100.00	$ 205.10
19	$ 205.10	$ 3.08	$ 208.17	$ 100.00	$ 108.17
20	$ 108.17	$ 1.62	$ 109.80	$ 100.00	$ 9.80
21	$ 9.80	$ 0.15	$ 9.95	$ 9.95	$ -0-
			TOTAL	$2,009.95	

That's a difference of $260 ($2,010 minus $1,750). You were also enticed into making the purchase because of the sale markdown from $1,925. As you can see, however, by putting the golf clubs on your credit card, you have ended up paying more for the clubs that you would have paid at full price. Would you have bought the clubs if you knew they actually would cost you $2,010?

While this example isolated the golf club purchase, in reality, you might well have been using the credit card for other purchases as you were paying off the clubs. Even if you paid off these items in full, you would have probably been charged interest on them because you were keeping a balance on your card. This interest would further increase the price of the clubs.

The Problem with Paying Just the Minimum Balance

Often when you're in a financial bind you can be tempted to pay only the minimum balance on your credit card. That's a risky proposition, since interest is accruing on your unpaid balance and probably on any new purchases as well.

Let's consider the golf clubs purchase above again. Let's assume that you are no longer using this card for new purchases, and you are paying the minimum $30 balance each month. How long will it take you pay off the golf clubs, and how much did they actually cost you? I have done the math for you using a spreadsheet program on the computer, because it would have been too unwieldy to tackle by hand.

After five years of minimum payments on the $1,750 golf clubs, you would still have a balance of $1,336.00. To completely pay off the clubs, it would take you over eleven years. (You might have given up golf in favor of bowling by then.) And, the clubs would have cost you almost $4,000.00, more than double their original price.

You can see by this example how difficult it is to get out of the cycle of credit-card debt once you are in it. You can imagine that, if the debt were up in the tens of thousands, it would be virtually impossible to pay off the card completely using the minimum payment. And the longer it takes you pay off the purchase, the more it will cost you to buy a specific item.

Don't despair, though. Help is on its way.

Paying Them Off

Okay, what happens when you're stuck with a lot of credit-card debt? You have learned your lesson. You have torn up all your credit cards and have sworn to pay only with cash in the future. You are still stuck with enormous debt. How do you best pay it off?

In general, you should do the following things:

1. Stop charging on any cards that have a balance that you cannot pay off every month.
2. Pay off the cards with the highest interest rate first.
3. Look into other kinds of conventional loans to pay off your cards, that have less interest than your cards are charging. Use the techniques described in the "Calculating Monthly Payments" section of this chapter to evaluate different loans.
4. If you are so deep in credit-card debt that you cannot see a way out, do not despair. You are not alone. Look in the phone book or on the Internet for one of the many reputable private, government, and not-for-profit organizations that will help you consolidate your loans and make a plan to get out of debt. Good luck!

The following example offers one strategy.

PROBLEM

You have racked up $13,500 of credit-card debt. Your card is charging 16% annual interest. You are no longer using the card, so no new purchases are coming in. Your sister offers to lend you the $13,500 to pay off your debt. She is asking for 7% interest. She wants the loan paid off in full in five years. You know your sister's 7% loan will be a better deal for you than continuing to pay off your 16%-

APR card. Before you commit to it, however, you want to know what the monthly payments would be.

STEP 1

Consider your sister's loan proposal. List the factors you will need to input into a financial calculator.

amount of loan = $13,500

annual interest rate = 7%

number of years to pay off the loan = 5

STEP 2

Input the factors in Step 1 into a financial calculator, and ask it to determine your monthly payment.

monthly payment = $267

SOLUTION

If you can manage the payments, swallow your pride and accept your sister's offer.

If you have accumulated a large balance on your credit card and want to begin paying it off, stop using it for new purchases so you will not be charged interest on them. If you absolutely must use a credit card for new purchases, get one that requires you to pay off the balance every month.

In a Nutshell

CREDIT CARDS

1. Credit cards vary significantly when it comes to interest rate, finance charges, fees, etc. Shop around for a card that suits your situation. Beware of low introductory interest rates that increase quickly.

continued

continued from previous page

2. If you do not pay off your balance every month, buying with a credit card is one of the most expensive ways to buy because of the high amount of interest charged not only on your balance, but on new purchases as well.

3. Think about the real price of something bought with your credit card if you do not pay off the balance each month. The real price is equal to the total amount of the payments you'll make on your card to cover that purchase. Ask yourself if you would have bought the item at the real price.

4. Paying only the minimum balance further increases the cost of buying something and makes it increasingly difficult to get out of debt.

5. If you are trying to reduce your credit-card debt, cut up your cards. Make no new purchases on cards that carry a balance, since you probably will be charged interest on new purchases before you have a chance to be billed for them. If you absolutely need a credit card, get one that requires you to pay off the balance every month.

6. Because of the high rate of interest on credit cards, look into other kinds of loans to pay off your cards.

7. If you are so deep in credit-card debt that you cannot see a way out, do not despair. You are not alone. Look in the phone book or on the Internet for one of the many reputable government or not-for-profit organizations that will help you make a plan.

 You can no longer deduct interest from credit cards and personal loans from your taxes.

Considering Taxes in Your Loan Decisions

Now that you are an expert on borrowing money, you might want to take your expertise a step farther by considering taxes in your loan decisions. Like investments, loans should be considered *after-tax* as well, if the interest on the loan is tax-deductible. While deducting loan interest for tax purposes is more limited than it once was, interest on most home mortgages and home-equity loans usually can still be deducted.

 Interest on second homes is not always deductible. It is best to check with your accountant.

If you are able to deduct your loan's interest from your taxes, it actually costs you less to borrow than the stated interest rate. To compute the true *after-tax interest rate*, you use a similar process to the one you used in considering investments in chapter 10. Follow these steps:

1. Look at the Estimating Your Marginal Tax Rate table on page 205 to estimate your marginal tax rate.
2. Change your marginal tax rate into a decimal by dividing it by 100.
3. Calculate the multiplier:
   multiplier = 1 minus marginal tax rate
4. Calculate the after-tax interest rate:
   after-tax interest rate = pretax interest rate times multiplier

Let's look at an example of how to calculate the after-tax interest rate on a loan and how to compare investments based on that knowledge.

PROBLEM

You have an extra $15,000 to invest. You and your husband make a combined $200,000, and you file jointly. You are trying to decide whether it makes more sense to pay down your 8% mortgage with the $15,000 or to invest the $15,000 in the stock market where you expect to make 7% over time. You have used the techniques in chapter 10, "Investment Basics," to determine that your after-tax

rate of return on the stocks is 5.6%. What is the after-tax interest rate on continuing to borrow $15,000 at 8%? And which is the better investment decision?

█ STEP 1

Look in the Estimating Your Marginal Tax Rate table on page 205 to estimate your marginal tax rate based on your $200,000 combined income and the fact that you are married and file your taxes jointly.

35%

█ STEP 2

Change your marginal tax rate from Step 1 (35%) into a decimal by dividing it by 100.

35% = 35 divided by 100 = .35

█ STEP 3

Calculate the multiplier by subtracting the marginal tax rate in decimal form from Step 2 (.35) from 1.

multiplier = 1 – .35 = .65

█ STEP 4

Calculate the after-tax interest rate on the mortgage by multiplying the pretax rate (8%) by the multiplier from Step 3 (.65)

after-tax interest = 8% times .65 = 5.2%

█ STEP 5

Compare the after-tax interest rate to continue to borrow the $15,000 through your mortgage against the after-tax rate of the return you get on the $15,000 in the stock market.

after-tax return on stock = 5.6%

after-tax interest rate on mortgage = 5.2%

■ SOLUTION

Put the money in the stock market, since you are hoping to get more of a return there than it costs you to borrow.

Notice that the amount of the loan did not matter. You just need to compare interest rates.

Investing is a very personal decision. Sometimes, even if the numbers point in a certain direction, you would just feel better doing something else. Some people get a lot of comfort paying down their mortgage, even if it is not the best financial use of their money. Those subjective factors are important and should be integrated with the numbers side of investing.

In a Nutshell

CONSIDERING TAXES IN YOUR LOAN DECISIONS
Calculating After-Tax Interest Rate on Loans

1. Look at the Estimating Your Marginal Tax Rate table on page 205 to determine your estimated marginal tax rate.
2. Change your marginal tax rate into a decimal by dividing it by 100.
3. Calculate the multiplier:
   multiplier = 1 minus marginal tax rate
4. Calculate the after-tax interest rate:
   after-tax interest rate = pretax interest rate times multiplier

---

Borrowing money is a two-sided affair. It can help make dreams come true. It can also cause nightmares. The math basics in this chapter should help you be more aware of what is involved with buying on credit, so that you can be a more savvy and knowledgeable consumer and investor.

Appendix 1
Twelve Tips to Stress-Free Math

1. Follow your instincts. They were around long before mathematicians got into the act.
2. Don't sweat the small stuff. If the problem you are confronting does not require a lot of accuracy, just estimate your answer.
3. Give your answer a reality check. If you do a mortgage calculation, and it tells you that your monthly payments on a $500,000 house would be $25 a month, chances are you made a mistake somewhere.
4. Only simplify fractions if you have to. For example, if you double a recipe, and your calculation tells you that you need ⅓ cup of sugar, measure out four ⅓ cups if it's quicker than changing ⅔ into 1⅓.
5. If you cannot figure out something in your head, don't sweat it. Use your calculator whenever you can.
6. Remember that a number without a visible decimal point really has one at the end of the number. For example, 375 is the same as 375.0.
7. If you are multiplying a whole number (that is a number that is not a fraction or does not have anything after the decimal) by a power of 10, just multiply it by the multiple without the zeroes, then add the zeroes to the answer. For example, to multiply 25 by 3,000, first muiltiple 25 by 3 (= 75), then just add three zeroes to the end (since 3,000 has three zeroes) to get 75,000.

8. To divide by a power of 10 (10, 100, 1,000, etc.), just move the decimal place to the left the number of zeroes in the divider. For example, to divide 37.5 by 100, just move the decimal two places to the left (since 100 has two zeroes) to get .375.

9. If your kids want help with their math homework, and you don't have a clue, don't fess up. Just tell them they will learn more if they work it out by themselves. Keep asking leading questions, like "What would you do next?" Between the two of you, you will probably figure it out.

10. A percentage is just a number divided by 100: 25% equals .25; 50% equals .50.

11. When calculating area or volume, make sure all your measurements are in the same unit (like inches, feet, or yards).

12. And don't forget—if you get stuck—you can always use your fingers, and use your toes.

APPENDIX 2
Fractions

The top of a fraction is commonly called the "numerator." The bottom of a fraction is commonly called the "denominator." Here, we will just use "top" and "bottom" to keep it simple.

Multiplying Fractions

To multiply a fraction by another fraction, first multiply the top of the first fraction by the top of the second fraction. Then multiply the bottom of the first fraction by the bottom of the second fraction.

Example: $\dfrac{2}{5}$ times $\dfrac{3}{7}$ = ?

Multiply the tops of the fractions by one another and the bottoms by one another.

$$\frac{2}{5} \text{ times } \frac{3}{7} = \frac{2 \text{ times } 3 = 6}{5 \text{ times } 7 = 35}$$

Solution: $\dfrac{6}{35}$

When multiplying a fraction by a whole number (a number that is not a fraction) turn the whole number into a fraction by making the top of the new fraction the number itself and making the bottom of the fraction "1." For example, $4 = \dfrac{4}{1}$. Then substitute the new fraction in for the whole number and multiply as in the steps above.

Example: 3 times $\dfrac{2}{5}$ = ?

Substitute $\frac{3}{1}$ for 3 and multiply as above.

$$3 = \frac{3}{1}$$

$$3 \text{ times } \frac{2}{5} = \frac{3}{1} \text{ times } \frac{2}{5} = \frac{3 \text{ times } 2}{1 \text{ times } 5} = \frac{6}{5}$$

Solution: $\frac{6}{5}$

Dividing Fractions

To divide one fraction by another, first flip the second fraction upside down. A flipped fraction is called a "reciprocal." For example, the reciprocal of $\frac{6}{7} = \frac{7}{6}$. Then multiply the first fraction by the reciprocal of the second fraction.

Example: $\frac{4}{9}$ divided by $\frac{1}{4}$ = ?

1. Find the reciprocal of the second fraction.

   The reciprocal of $\frac{1}{4} = \frac{4}{1}$.

2. Multiply the first fraction by the result of #1 as above.

   $$\frac{4}{9} \text{ divided by } \frac{1}{4} = \frac{4}{9} \text{ times } \frac{4}{1} = \frac{4 \text{ times } 4}{9 \text{ times } 1} = \frac{16}{9}$$

Solution: $\frac{16}{9}$

Adding Fractions

If the bottoms are the same:

Add the tops and keep the bottom the same.

Example: $\frac{3}{7}$ plus $\frac{2}{7}$ = ?

$$\frac{3}{7} \text{ plus } \frac{2}{7} = \frac{(3 \text{ plus } 2)}{7} = \frac{5}{7}$$

Solution: $\dfrac{5}{7}$

If the bottoms are different:

You need to get both fractions into a form where they have the same bottoms. This is called finding a "common denominator." To do this, you first need to list the multiples of each bottom until you find a multiple that is the same, or common, to each. A "multiple" is just a number multiplied by 1, 2, 3, etc. So, multiples of 3 would be 3 (3 times 1), 6 (3 times 2), 9 (3 times 3), 12 (3 times 4), etc. Once you determine the multiples for each fraction bottom, find the smallest one that is common to both fractions. Then multiply both the top and bottom of each fraction by a number that will, when multiplied by the bottom, produce the common denominator. Substitute these new fractions for the old ones. Since both new fractions now have the same bottom, you can just add them as you did above.

Example: $\dfrac{2}{3}$ **plus** $\dfrac{1}{4}$ = ?

1. List the multiples for the bottom of each fraction. It helps to start with the smallest bottom. You can stop forming multiples with the second fraction when you find a match with the first.

   For $\dfrac{2}{3}$, you find multiples of 3: 3 (1 times 3), 6 (2 times 3), 9 (3 times 3), 12 (4 times 3), 15 (5 times 3), 18 (6 times 3) . . .

   For $\dfrac{1}{4}$, you find multiples of 4: 4 (1 times 4), 8 (2 times 4), 12 (3 times 4) . . .

2. Find the smallest multiple that is common to both fractions. That number will be the new bottom of both fractions.

   12

3. Look at #1 to see what number you had to multiply each bottom by to get the common multiple in #2.

For $\frac{2}{3}$, you had to multiply 3 by $\underline{4}$ to get the common multiple of 12.

For $\frac{1}{4}$, you had to multiply 4 by $\underline{3}$ to get the common multiple of 12.

4. Multiply the top and bottom of each fraction by the underlined multiplier from #3. (Multiplying the top and bottom of a fraction by the same number does not change its value.)

$$\frac{2}{3} = \frac{2 \text{ times } 4}{3 \text{ times } 4} = \frac{8}{12}$$

$$\frac{1}{4} = \frac{1 \text{ times } 3}{4 \text{ times } 3} = \frac{3}{12}$$

5. Now substitute the new forms of each fraction. Because both fractions now have the same bottom, you can just add the tops together and leave the bottom alone.

$$\frac{2}{3} \text{ plus } \frac{1}{4} = \frac{8}{12} \text{ plus } \frac{3}{12} = \frac{8 \text{ plus } 3}{12} = \frac{11}{12}$$

Solution: $\frac{11}{12}$

Subtracting Fractions

If the bottoms are the same:

Subtract the tops, and keep the bottom the same.

Example: $\frac{5}{8}$ minus $\frac{2}{8}$ = ?

$$\frac{5}{8} \text{ minus } \frac{2}{8} = \frac{5 \text{ minus } 2}{8} = \frac{3}{8}$$

Solution: $\frac{3}{8}$

If the bottoms are different:

As in adding fractions above, to subtract fractions with different bottoms, both fractions need to be changed so that they have the same bottom (see "Adding Fractions," above). After the fractions are in a form where they have the same bottom, you can just subtract them as you did above.

Example: $\dfrac{4}{5} - \dfrac{2}{15} = ?$

1. List the multiples of the bottom number of each fraction. It helps to start with the fraction with smallest number on the bottom. You can stop forming multiples for the second fraction when you find a match with the first.

   For $\dfrac{4}{5}$, you find multiples of 5: 5 (1 times 5), 10 (2 times 5), 15 (3 times 5)

   For $\dfrac{2}{15}$, you find multiples of 15: 15 (1 times 15)

   In this example, the bottom of the second fraction is actually already a multiple of the first.

2. Find the smallest multiple that is common to both fractions. That number will become the new bottom of both fractions.

   15

3. Look at #1 to see what number you had to multiply each bottom by to get the common multiple in #2.

   For $\dfrac{4}{5}$, you had to multiply 5 by 3 to get the common multiple of 15.

   For $\dfrac{2}{15}$, you had to multiply 15 by 1 to get the common multiple of 15.

4. Multiply the top and bottom of each fraction by the underlined multiplier from #3.

   $$\frac{4}{5} = \frac{4 \text{ times } 3}{5 \text{ times } 3} = \frac{12}{15}$$

$$\frac{2}{15} = \frac{2 \text{ times } 1}{15 \text{ times } 1} = \frac{2}{15}$$

As you can see, any number multiplied by 1 is just that number $\left(\frac{2}{15} \text{ times } \frac{1}{1} = \frac{2}{15}\right)$, so you just leave $\frac{2}{15}$ as it is.

5. Now substitute the new forms of each fraction. Because both fractions now have the same bottom, you can just subtract the tops and leave the bottom alone.

$$\frac{4}{5} \text{ minus } \frac{2}{15} = \frac{12}{15} \text{ minus } \frac{2}{15} = \frac{12-2}{15} = \frac{10}{15}$$

Solution: $\frac{10}{15}$

Technically, this fraction is not in its "simplest form." See the next section for simplifying fractions.

Simplifying Fractions

A fraction is in its "simplest form" when you can no longer divide the top and bottom evenly (without anything leftover) by any number except for 1. The process is trial and error. Just keep dividing the top and bottom by the same number until there is no number that can go into the top and bottom of the fraction evenly. It does not matter what numbers you try or the order in which you try them.

Example: Put $\frac{6}{12}$ into its simplest form.

1. Divide the top of the fraction and the bottom of the fraction by 2.

$$\frac{6}{12} = \frac{6 \text{ divided by } 2}{12 \text{ divided by } 2} = \frac{3}{6}$$

2. Divide the top of the new form of the fraction from #1 by 3.

$$\frac{3}{6} = \frac{3 \text{ divided by } 3}{6 \text{ divided by } 3} = \frac{1}{2}$$

3. Since no number, except 1, can be divided evenly into both 1 and 2, you are finished.

Solution: $\dfrac{1}{2}$

To make the simplifying process move more quickly, try to divide the top and the bottom by the largest number possible. In the example above, if you had started by dividing both the top and bottom of $\dfrac{6}{12}$ by 6, you would have more quickly gotten to

$$\frac{6 \text{ divided by } 6}{12 \text{ divided by } 6} = \frac{1}{2}.$$

Changing Fractions into Mixed Numbers

Sometimes you'll want to change a fraction into a mixed number. A mixed number consists of a whole number (a number that is not a fraction) and a fraction. To calculate the whole number part of the mixed number, divide the top by the bottom and save the remainder. The fraction part of the mixed number will consist of the remainder over the original fraction bottom.

Example: Change $^{35}\!/_3$ into a mixed number.

1. Divide the top by the bottom.

   35 divided by 3 = 11, with 2 left over (called the "remainder").

2. The whole number part of the mixed number is the first result of #1 (the one that is *not* left over).

   11

3. The fraction part of the mixed number is the leftover part from #1 over the bottom of the original fraction.

   $^2\!/_3$

4. The mixed number becomes the result of #2 plus the result of #3.

   $11^2\!/_3$

Solution: $11^2\!/_3$

Changing Mixed Numbers into Fractions

To change a mixed number (a whole number and a fraction) into a fraction, first find the top of the new fraction by multiplying the bottom of the fraction by the whole number and adding to it the top of the fraction. Put that number over the original fraction bottom to form the final fraction.

Example: Change 5⅞ to a fraction.

1. Multiply the bottom of the fraction by the whole number.
   8 times 5 = 40
2. Add the result of #1 and the top of the fraction.
   40 plus 7 = 47
3. The top of the new fraction is the result of #2. The bottom of the new fraction is the bottom of the original fraction.
   ⁴⁷⁄₈

Solution: ⁴⁷⁄₈

Changing Fractions to Decimals

To change fractions to decimals with your calculator, just divide the top of the fraction by the bottom.

Example: ¼ = ?

Divide the top of the fraction by the bottom.

1 divided by 4 = .25

Solution: .25

For common fractions, you can also just look at the Table of Common Fractions in Decimal Form below to find the decimal equivalent.

Changing Decimals to Fractions

There is not an easy way to precisely change a decimal into a fraction. However, you can use the chart below to get a pretty close estimate. In the second column, find the decimal that is the closest to the one you want to change to a fraction.

Table of Common Fractions in Decimal Form	
Fraction	**Decimal**
1/20	.05
1/10	.10
1/8	.125
1/5	.20
1/4	.25
1/3	.333
3/8	.375
2/5	.40
1/2	.50
3/5	.60
5/8	.625
2/3	.67
7/10	.70
3/4	75
4/5	.80
7/8	.875
9/10	.90

Appendix 3
Percents

What Is a Certain Percentage of a Number?

First convert the percent to a decimal by dividing it by 100. Then multiply that decimal by the number.

Example: What is 35% of 195?

1. Convert the percentage to a decimal value by dividing it by 100.

   35% = .35

2. Multiply the decimal from #1 by the total.

   35 times 195 = 68.25

Solution: 68.25

To divide by 100 when changing a percent into a decimal, just place a decimal point at the end of the percentage, and then move the decimal place two digits to the left. If there are not two digits to the left, add a zero in front of the one that is present, *then* move over two digits. For example, 7% = 07% = .07.

What Percentage Is One Number of Another?

Divide the first number by the second. Then multiply that result by 100 and add a percent sign (%).

Example: What percentage is 17 of 240?

1. Divide the first number by the second.

   17 divided by 240 = .071

2. Multiply the result of #1 by 100 and add a percent sign.

   .071 times 100 = 7.1%

Solution: 7.1%

When multiplying by 100 to form a percent, just move the decimal point two digits to the right. If there are not two digits, add a zero after the last digit, then move. For example, .6 = .60 = 60%.

What is the Percentage Change from One Number to Another Number?

Subtract the old number from the new number. Divide that result by the old number. Multiply by 100 and add a percent sign. Some people remember this as "new minus old over old." If you end up with a positive percentage change, that means the change was an increase. If you end up with a negative percentage change, do not panic. That just means the change was a decrease.

Example: What is the percentage change from 335 to 392?

1. Subtract the old number from the new number.

   392 minus 335 = 57

2. Divide the result of #1 by the old number.

   57 divided by 335 = .17

3. Multiply the result of #2 by 100 and add a percent sign.

   .17 times 100 = 17%

Solution: 17% *increase*

Example: What is the percentage change from 112 to 85?

1. Subtract the old number from the new number.

   85 minus 112 = −27

2. Divide the result of #1 by the old number.

   −27 divided by 112 = −.24

3. Multiply the result of #2 by 100 and add a percent sign.

   −.24 times 100 = −24%

Solution: 24% *decrease*

Changing Fractions into Percents

Divide the top of the fraction by the bottom, multiply by 100 and add a percent sign.

Example: What percent is ¾?

1. Divide the top of the fraction ("numerator") by the bottom ("denominator").

    3 divided by 4 = .75

2. Multiply the result of #1 by 100 and add a percent sign.

    .75 times 100 = 75%

Solution: 75%

The table below provides a list of some common fractions and their equivalents in percents. These equivalents are helpful to know, since you might not always have your calculator with you.

Table of Common Fractions in Percent Form	
Fraction	Percent
¼	25%
⅓	33.3%
½	50%
⅔	66.7%
¾	75%
1	100%

Changing Percentages into Decimals

Just divide the percentage by 100 and drop the percent sign.

Example: What is the decimal form of 63%?

Divide the percent by 100 and drop the percent sign.

63% divided by 100 = .63

Solution: .63

Appendix 4
Rounding

Places

To understand rounding, you first need to understand that every number has the same structure, and each digit in the number has a certain specific place in that structure. For example, in the number 1,237, the 1 is in the thousands place, the 2 is in the hundreds place, the 3 is in the tens place, and the 7 is in the ones place. So, the number 1,237 is actually:

(1 times 1000) plus (2 times 100) plus (3 times 10) plus (7 times 1), which equals 1,000 plus 200 plus 30 plus 7 (= 1,237)

Look at the illustration below to see how places are arranged.

millions hundred   ten   thousands hundreds tens   ones   tenths   hun-   thou-
          thousands thousands   dredths   sandths

Rounding Whole Numbers

Draw a line under the place to which you are rounding (e.g., thousands, hundreds, etc.). Look at the number directly to the right of the underlined number. If the result is 5 or over, round *up* by increasing the underlined number by 1. If the result is less than 5, round *down* by leaving the underlined number alone. If the underlined number is 9 and you are rounding up, then replace the underlined number with 0, and increase the number to its left by 1. Fill the places to the right of the underlined number with zeros.

Example: Round 135,649 to the nearest hundred.

1. Draw a line under the place to which you are rounding.

    135,<u>6</u>49

2. Look at the number to the right of the underlined number.

    4

3. If the result of #2 is 5 or over, round *up* by increasing the underlined number by 1. If the result of #2 is less than 5, round *down* by leaving the underlined number alone. If the underlined number is 9 and you are rounding up, then replace the underlined number with 0, and increase the number to its left by 1. Fill the places to the right of the underlined number with zeros.

    135,<u>6</u>00

Solution: 135,600

Example: Round 63,498 to the nearest ten.

1. Draw a line under the place to which you are rounding.

    63,4<u>9</u>8

2. Look at the number to the right of the underlined number.

    8

3. If the result of #2 is 5 or over, round *up* by increasing the underlined number by 1. If the result of #2 is less than 5, round *down* by leaving the underlined number alone. If the underlined number is 9 and you are rounding up, then replace the underlined number with 0, and increase the number to its left by 1. Leave the space to the right of the underlined number blank.

    63,5<u>0</u>0

Solution: 63,500

Rounding Decimals

To round decimals, you use the same rounding techniques as described above.

Example: Round 12.735 to the nearest tenth.

1. Draw a line under the place to which you are rounding.

   12.<u>7</u>35

2. Look at the number to the right of the underlined number.

   3

3. If the result of #2 is 5 or over, round *up* by increasing the underlined number by 1. If the result of #2 is less than 5, round *down* by leaving the underlined number alone. If the underlined number is 9 and you are rounding up, then replace the underlined number with 0, and increase the number to its left by 1. Leave the space to the right of the underlined number blank.

   12.<u>7</u>

Solution: 12.7

Appendix 5
Perimeter, Area, and Volume

In calculating perimeter, area, and volume, make sure that all the measurements you are using for one problem are the same. For example, if you are calculating the volume of a box, the height, width, and depth should all be in inches or feet or some other equivalent measure. If they are not, consult appendix 6, "Changing Units of Measure."

Perimeter (The Distance around an Object)

Consult the table on page 258 for specific formulas.

Examples

Square
Example: What is the perimeter of a square 5 feet wide?
Multiply the width of one side by 4.

4 times 5 feet = 20 feet

Solution: 20 feet

Rectangle
Example: What is the perimeter of a rectangle 3 miles wide and 7 miles long?

1. Multiply the width by 2.

   2 times 3 miles = 6 miles

2. Multiply the length by 2.

   2 times 7 miles = 14 miles

Perimeter

Square

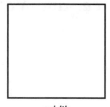

width
perimeter = 4 times width

Rectangle

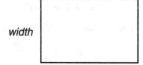

length
perimeter = 2 times width
plus 2 times length

Circle

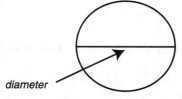

perimeter = 3.14 times diameter

Triangle

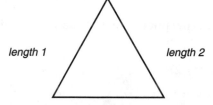

length 3
perimeter = length 1 plus length 2 plus length 3

3. Add the results of #1 and #2.

6 miles plus 14 miles = 20 miles

Solution: 20 miles

Circle

Example: What is the perimeter of a circle that has a diameter of 6 inches?

Multiply the diameter of the circle by 3.14.

3.14 times 6 inches = 18.84 inches

Solution: 18.84 inches

Triangle

Example: What is the perimeter of a triangle with sides of 75 yards, 63 yards, and 25 yards?

Add 75 yards plus 63 yards plus 25 yards.

75 yards plus 63 yards plus 25 yards = 163 yards

Solution: 163 yards

Area

Area is measured in "squares" such as square acres, square feet, square yards, etc. Consult the table on page 260 for specific formulas.

Examples

Square

Example: What is the area of a square 40 miles wide?

Multiply the width of the square by itself.

40 miles times 40 miles = 1,600 square miles

Solution: 1,600 square miles

Area

Square

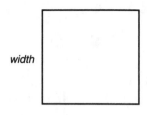

area = width times width

Rectangle

area = width times length

Circle

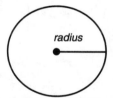

area = 3.14 times radius times radius

Triangle

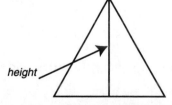

area = 1/2 times base times height

Rectangle

Example: What is the area of a rectangle that is 6 meters wide and 9 meters long?

Multiply the width of the rectangle by its length.

6 meters times 9 meters = 54 square meters

Solution: 54 square meters

Circle

Example: What is the area of a circle that has a radius of 5 yards?

1. Multiply the length of the radius by the length of the radius.

    5 yards times 5 yards = 25 square yards

2. Multiply the result of #1 by 3.14.

    25 square yards times 3.14 = 78.5 square yards

Solution: 78.5 square yards

Triangle

Example: What is the area of a triangle that has a base of 8 inches and a height of 11 inches

1. Divide the length of the base of the triangle by 2.

    8 inches divided by 2 = 4 inches

2. Multiply the result of #1 by the height of the triangle.

    4 inches times 11 inches = 44 square inches

Solution: 44 square inches

Volume

Volume is represented in cubic measurements, such as cubic inches, cubic feet, etc. For example, one cubic foot would be the volume of an ice cube that is one foot wide, one foot long, and one foot tall. Consult the table on page 262 for specific formulas.

Volume

Cube

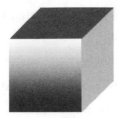

width
volume = width times width times width

Box

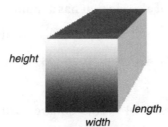

height

length

width
volume = width times length times height

Cylinder

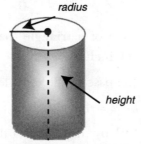

radius

height

volume = 3.14 times radius times radius times height

Cone

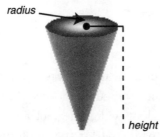

radius

height

volume = 1.05 times radius times radius times height

Examples

Cube

Example: What is the volume of a cube that is 7 feet wide?

Multiply the width of the cube times itself twice.

7 feet times 7 feet times 7 feet = 343 cubic feet

Solution: 343 cubic feet

Box

Example: What is the volume of a box that is 8 inches wide, 12 inches long, and 4 inches high?

1. Multiply the width of the box by its length.

   8 inches times 12 inches = 96 square inches

   (This is the area of the bottom of the box.)

2. Multiply the result of #1 by the height of the box.

   96 square inches times 4 inches = 384 cubic inches

Solution: 384 cubic inches

Cylinder

Example: What is the volume of a cylinder that has a radius of 15 feet and a height of 40 feet?

1. Multiply the radius by itself.

   15 feet times 15 feet = 225 square feet

2. Multiply the result of #1 by 3.14.

   225 square feet times 3.14 = 706.5 square feet

   (This is the area of the bottom of the cylinder.)

3. Multiply the result of #2 by the height of the cylinder.

   706.5 square feet times 40 square feet = 28,260 cubic feet

Solution: 28,260 cubic feet

Cone

Example: What is the volume of a cone that is 8 centimeters high and has a radius of 3 centimeters at the widest point?

1. Multiply the radius by itself.

   3 centimeters times 3 centimeters = 9 square centimeters

2. Multiply the result of #1 by 1.05.

   9 square centimeters times 1.05 = 9.45 square centimeters

3. Multiply the result of #2 by the height.

   9.45 square centimeters times 8 centimeters = 75.6 cubic centimeters

Solution: 75.6 cubic centimeters

APPENDIX 6
Changing Units of Measure

To change from one unit of measure to another, just follow the directions on the chart below. This chart combines the information from all the measurement changing tables in the book.

To go from . . .	to . . .	you . . .	. . . . .	by . . .
centimeters	inches	divide	# of centimeters	2.54
¼ cups	tablespoons	multiply	# of ¼ cups	4
cubic feet	cubic yards	divide	# of cubic feet	27
cubic yards	cubic feet	multiply	# of cubic yards	27
cups	tablespoons	multiply	# of cups	16
cups	ounces	multiply	# of cups	8
cups	pints	divide	# of cups	2
cups	quarts	divide	# of cups	4
feet	inches	multiply	# of feet	12
feet	yards	divide	# of feet	3
feet	meters	divide	# of feet	3.28
gallons	quarts	multiply	# of gallons	4
gallons	liters	multiply	# of gallons	3.79
inches	feet	divide	# of inches	12
inches	yards	divide	# of inches	36
inches	centimeters	multiply	# of inches	2.54
kilograms	pounds	multiply	# of kilograms	2.2
kilometers	miles	multiply	# of kilometers	.625
liters	gallons	divide	# of liters	3.79
liters	pints	multiply	# of liters	2.11

continued

To go from ...	to ...	you ...	.....	by ...
liters	quarts	multiply	# of liters	1.06
liters	ounces	multiply	# of liters	33.81
meters	feet	multiply	# of meters	3.28
meters	yards	multiply	# of meters	1.09
miles	feet	multiply	# of miles	5280
miles	kilometers	multiply	# of miles	1.6
ounces	cups	divide	# of ounces	8
ounces	pounds	divide	# of ounces	16
pints	cups	multiply	# of pints	2
pints	quarts	divide	# of pints	2
pints	liters	divide	# of pints	2.11
pounds	ounces	multiply	# of pounds	16
pounds	kilograms	divide	# of pounds	2.2
quarts	cups	multiply	# of quarts	4
quarts	gallons	divide	# of quarts	4
quarts	pints	multiply	# of quarts	2
quarts	liters	divide	# of quarts	1.06
square feet	square inches	multiply	# of square feet	144
square feet	square yards	divide	# of square feet	9
square inches	square feet	divide	# of square inches	144
square yards	square feet	multiply	# of square yards	9
tablespoons	¼-cups	divide	# of tablespoons	4
tablespoons	cups	divide	# of tablespoons	16
tablespoons	teaspoons	multiply	# of tablespoons	3
teaspoons	tablespoons	divide	# of teaspoons	3
yards	feet	multiply	# of yards	3
yards	inches	multiply	# of yards	36
yards	meters	divide	# of yards	1.09

To go from Celsius temperature to Fahrenheit, multiply Celsius temperature by 1.8 and add 32 degrees.

To go from Fahrenheit temperature to Celsius, subtract 32 degrees from the Fahrenheit temperature and multiply that figure by .56.

Afterword

Well, you've done it. You've made your way through more math than you had ever dreamed you could—or would, for that matter. You might have read the whole book cover to cover. Chances are, however, that you have read only portions of it, picking it up when you needed it.

I hope *Use Your Fingers, Use Your Toes* has been helpful to you and has explained things in ways that you could understand. I hope that walking through example problems step-by-step has helped you solve the thorniest of your everyday math problems, and that the In a Nutshell boxes gave you quick reminders when you needed it.

But mostly, I hope that as you have read the book your belief in your own ability to handle everyday math has grown and grown. Did you notice as you were going through a particular example that much of the math sounded familiar? Did you find yourself saying, "Oh yeah, I remember now"?

I firmly believe that most of what you need to solve everyday math problems is already inside your head somewhere. Maybe it's a little jumbled, but it is definitely there. Somewhere along the line you just lost faith in your own math skills. Then, you began to panic. We all have moments like that. Mine are in elevators. (But that's another book.)

The wonderful thing is that this little book will always be around to boost your confidence when you need it. Keep it someplace handy. If you need help, it will always be there for you.

Index